Plastics and Sustainability

Scrivener Publishing
100 Cummings Center, Suite 541J
Beverly, MA 01915-6106

Publishers at Scrivener
Martin Scrivener (martin@scrivenerpublishing.com)
Phillip Carmical (pcarmical@scrivenerpublishing.com)

Plastics and Sustainability

2nd Edition

Grey is the New Green:
Exploring the Nuances and
Complexities of Modern Plastics

Michael Tolinski
and
Conor P. Carlin

Scrivener
Publishing

WILEY

Library of Congress Cataloging-in-Publication Data

ISBN 978-1-119-59184-9

Cover image: Top right: Recycled PET (rPET) clamshell with multiple recycling logos. Image courtesy of C. Carlin.
Top left: Nespresso-compatible coffee capsules by Georg Menshen. Image courtesy of C. Carlin.
Center: Adjustable prosthetic pylon manufactured and patented by Nonspec, Inc. attached to a 3D printed foot by Mercuris.
Cover design by Russell Richardson

Set in size of 11pt and Minion Pro by Manila Typesetting Company, Makati, Philippines

Contents

Acknowledgements

Mike Tolinski's first edition of *Plastics and Sustainability* was a uniquely balanced book that presented a complex topic in an objective way. With his background in materials science, engineering, and industry journalism, Mike had both insight and foresight when writing about polymers and their use in modern life. My efforts are mere updates to his foundational work.

I am grateful to many collaborators from across plastics, environmental, and government groups who provided current data, especially the research team at *Plastics News*, and my colleagues from the ranks of the Society of Plastics Engineers.

I wish to thank Martin Scrivener for his willingness to entertain a second edition of this book.

Notes on the 2nd Edition

A lot has happened in the world of plastics since 2010. Chemistry, however, moves more slowly. The majority of material related to polymer properties and general organic chemistry remains unchanged in this edition. There are exceptions for updated facts and figures related to production volumes and developments in bio-based polymers. There are many companies that have come and gone since 2010, most of which were start-ups working on novel uses of bio-based materials. In true Schumpeterian fashion, however, there are probably just as many new ones funded and operating today. Generally speaking, the level of public awareness of broad sustainability issues is higher now than in 2010, so some basic terms, definitions, and explanations have been removed as I assume a more educated reader today. Mike made extensive use of endnotes, many of which were taken from industry sources and proceedings at the time of writing. I have tried to update as many of these as possible, though some have been removed where the primary text has changed.

The book is relatively US-centric, though attempts are made to introduce facts, data, and developments from other regions of the world. Europe, in particular, has a very different regulatory system than the US, which has led to divergent policies and subsequent industrial evolution. Mike did not use a lot of visual elements (graphs, charts, illustrations, etc.) in the first edition. Because I want to expand the audience for this book, I have incorporated much more in the way of data visualization.

If it is true that the general public has a greater awareness of plastics-related topics today, this is still a large and complex global industry with many moving parts, not all of which can be seen outside of insider sources such as industry press, conference proceedings, and journal articles. Where it serves to enlighten a broader audience, these sources are used and referenced throughout the book. What may appear to be an obscure and uninteresting development in a research lab today can turn out to be the seed of a new product tomorrow.

Mike introduced each chapter with a fictional account of a plastics injection molding company navigating the changing world of polymer materials, especially bio-based plastics. I have chosen not to repeat or continue this story, but rather elected to insert new and current examples of commercial successes in new and environmentally friendly uses of (bio)plastics, where companies, products, and services have achieved market recognition. To attract a broader readership, and to acknowledge the increased awareness of the environmental impact of plastics, I have included several new and different perspectives from non-industry sources.

Preface

The June 2018 issue of *National Geographic* magazine featured a striking cover photo of what appeared to be an iceberg. It was, in fact, an image of plastic shopping bags mostly submerged in water. The title was "Planet or Plastic?" and the issue itself was the first installment of a multi-year examination of humanity's relationship with the "miracle material." Newspapers, websites, magazines, television reports, podcasts, and more were full of stories and images of littered plastic: marine life entangled with plastic fishing nets, rings, and straws; dead birds with stomachs full of microplastics; Southeast Asian rivers choked with single-use plastic sachets; Chinese, Indian, and Indonesian "waste pickers" atop mountains of polymer-based waste. Landmark reports, especially those published by the Ellen MacArthur Foundation, offered grim statistics related to plastics and the environment. One sensational finding in particular seemed to go viral, which is that if the current rate of plastics production and disposal continued, there would be more plastics than fish in the sea by 2050.* This meme was challenged and ultimately debunked, but it was too late. Thus began a torrent of public and private initiatives, multinational collaboratives, legislative bans, and, perhaps most importantly, an awakening of

---

* The exact text is: "*In a business-as-usual scenario, the ocean is expected to contain 1 tonne of plastic for every 3 tonnes of fish by 2025, and by 2050, more plastics than fish (by weight).*" World Economic Forum, Ellen MacArthur Foundation and McKinsey & Company, The New Plastics Economy — Rethinking the future of plastics (2016, http://www.ellen-macarthurfoundation.org/publications).

public awareness on the subject of plastic waste. The ubiquity, durability, and persistence of plastics were being questioned like never before.

Also in June 2018, *Plastics Engineering* magazine released their "Green Issue" which featured in-depth articles about new biodegradable and bioresorbable plastics; multimillion dollar commitments from major brands to improve the designs of plastic products for recycling or reuse; new programs in the US and Europe for more aggressive recycling and waste-to-energy targets; and new applications in automotive plastics for agricultural waste products, including rice hulls, coconut shells and other nontraditional fibers. In January 2019, a consortium of major global brands announced a $1.5bn initiative, "The Alliance to End Plastic Waste." This was the single biggest financial commitment to date from consumer-facing companies like P&G and Unilever, and plastics manufacturing firms like LyondellBasell and Dow Chemical, to address what has become a global phenomenon. Industry, it seemed, got the message; yet nagging skepticism remained in the face of continued plastic production and stagnating recycling data. Terms like "circular economy" and "new plastics economy" have supplanted "triple bottom line" as keystones in the attempts to address highly visible plastics litter.

Since the first edition of this book, published in 2010, there have been multiple seismic events in the world of plastics and sustainability that have altered the landscape dramatically. Though it is not the goal of this second edition to provide a comprehensive update on all industrial, legislative, or scientific events, I will attempt to categorize the most significant impacts and provide the most compelling examples. For example, on January 1, 2018, China put into effect its previously announced decision to ban imports of certain types of waste and recyclable materials. Dubbed "National Sword," this development fundamentally changed the way entire countries have to think about their own waste management. In May 2019, the Basel Convention produced a new international treaty designed to prevent the transfer of hazardous waste, including "mixed, unrecyclable, and contaminated plastic waste exports," from developed to less developed countries. The United States, which is not a party to the Basel Convention, nonetheless opposed the ratification of the agreement along with Argentina and Brazil. Other major developments will be covered along with their effects on everything from public policy, to private sector investments, to product design, all the way to how you and I must now think about not only what we buy and consume, but how we dispose of our purchases.

Technology is often cited as a savior to humanity's problems, yet our behavior is acknowledged as a significant contributor to environmental degradation. We are not rational actors. Will the promise of the circular economy adequately shift our behaviors without forcing sacrifice? Can we have our cake and eat it too? "Degradability is not a solution to littering," yet the behavior of polluters is slow to change. We are not paying for the damage we create in the meantime. In ten years, plastics waste recycling has increased by almost 80% in Europe, but global virgin resin capacity continues to increase in response to market demands. It's as if the circle is expanding even as we attempt to close it.

Do plastics actually threaten our future? If we look at greenhouse gas emissions, only 4-6% of crude oil is currently converted to plastics, though the figure is rising. In the context of the average western household, packaging represents a mere 1.7% of that unit's carbon footprint, with home heating, electricity usage, and transportation accounting for almost 50%. But litter or unmanaged plastic waste is an externality that is not covered by current GHG metrics or market-based pricing. Its visibility and ubiquity are forcing changes throughout industry and society, some of which are being felt by the consumer.

Yet the future is not doomed to look like the past. Forecasting is a mug's game, with experts often proven wrong in their predictions. In my business travels throughout the US, Europe, and Asia, I often attend regional seminars on plastics technology and the circular economy. Many authors and speakers acknowledge that materials, technologies, infrastructure, and regulations keep changing, though not always in concert. The pan-EU approach, for example, seeks to build a consensus that will allow the construction of an overarching framework within which all value chain players can work. The recently announced "Global Commitment" to eliminate plastic waste was signed by over 250 entities, but still only represents 20% of all plastic packaging produced globally. For now, the work done by the MacArthur Foundation offers both a vision and a strategy that has brought many disparate groups together. With growth set to continue in plastic packaging markets, the circular economy will have to grow with it.

More than ever, plastics are front and center in the public eye, for better or for worse, with a vigorous debate underway about how to deal with the challenges posed by the complex and amazing long-chain molecules we know as polymers.

January 2021
Plymouth, Massachusetts

1

General Introduction

This introductory chapter will address fundamental, current questions of interest related to plastics and sustainability. As with the other chapters in this book, this chapter will begin with a brief outline of its contents. The main content of each chapter provides details that similarly interested readers might consider when making sustainability decisions about plastics.

This chapter will define terms and arguments related to issues of environmental sustainability, restating arguments in ways that best highlight the challenges that plastics-consuming companies can focus their energies on. It will briefly introduce issues that are covered later in the book in more detail, also linking some well-known plastics controversies with the broader context of this book. Specifically, this chapter will:

- present an overview of the sometimes contradictory positive and negative features of polymer-based materials (1.1);
- illustrate consumers' dependence on plastics in their plastics-based lifestyles (1.2);
- provide a brief history of recent (and often controversial) sustainability issues concerning plastics (1.3);

Michael Tolinski and Conor P. Carlin. Plastics and Sustainability 2nd Edition: Grey is the New Green: Exploring the Nuances and Complexities of Modern Plastics, (1–28) © 2021 Scrivener Publishing LLC

- discuss the "need for green" – especially the social pressures that are forcing plastics manufacturers to take sustainability seriously (1.4); and
- provide an overview of the chapters in the remainder of the book (1.5).

The Growth of Plastics Literature

Published in 2011, *Plastic: A Toxic Love Story* by Susan Freinkel is a story about the ubiquity of plastics in daily life that inspired a bevy of similar books with dramatic titles and subtitles, e.g., *How to Give Up Plastic: A Guide to Changing the World, One Plastic Bottle at a Time, Life Without Plastic The Practical Step-by-Step Guide to Avoiding Plastic to Keep Your Family and the Planet Healthy*, and *F**k Plastic: 101 Ways to Free Yourself from Plastic and Save the World*. These books, and others, reflect a complex picture of the public's understanding of not only plastics, but their underlying chemical composition. The ecological and health impacts of plastics, their production, use, and disposal, are not always universally understood and in many cases are ripe for misunderstanding and even outright manipulation. When the internet delivers tailored results in a few clicks, the objective reader must work harder to find balance and ultimately be comfortable with ambiguity. Yet this is not an easy position to hold when human and animal health is at risk. Scientists can declaim Paracelsus' maxim[*] but also invoke the precautionary principle. Environmentalists can cite data about increasing plastics pollution in global waterways, but also acknowledge that life cycle analyses almost always favor lightweight polymer-based materials. Growth itself is central to the debate: can the earth's ecosystem continue to support humanity's needs or will we (have we) reached a tipping point where the carrying costs of natural capital have exceeded the benefits? One can easily move from a single point about banning plastic bags to the impact of climate change and encounter diametrically opposed views. Facts on the ground bear this out: plastics production has increased steadily since 1950 yet our ability to manage it effectively has not. What Freinkel and others represent is an acknowledgement that we face a dilemma when trying to balance the benefits and penalties of plastic.

---

[*]Sola dosis facit venenum – The dose makes the poison.

1.1   The Contradictions of Plastics

An honest assessment of plastics as being useful, important materials also requires admitting that plastic products' shapes, forms, compositions, uses, and material qualities are somewhat enigmatic or contradictory. Plastic products sometimes resemble objects from nature, although sometimes an alien, even science fiction form of nature. But normally they are not considered as being like anything that is natural; after all, plastic compositions are created by chemists and are mysterious to those who are not educated in organic chemistry.

The term "plastics" is itself inadequate and misleading in that it refers to a wide range of materials. Some plastics are rubbery and some do not melt when heated; some are strengthened with glass fiber to become composites, while others are used in the form of simple films or foams. Certain plastic products are weak, cheap, and disposable; others are strong and durable — yet all are labeled simply as plastic.

Other contradictions below relate to plastics' place in our economy and in our larger natural environment:

- Plastics are inexpensive… but their properties can be tailored for very high-value, engineering purposes.
- Plastics often have relatively simple chemical structures… but they can be extremely resistant to processes of natural decay, guaranteeing their long-term persistence in the environment.
- Plastic products are lightweight… but millions of tons of them have been consumed and discarded.
- Plastics are made from high-energy chemical feed-stock, becoming in essence "frozen fuel"… but plastic products and their inherent energy are commonly treated as waste for landfills, often after very brief lifetimes of use, unlike liquid or gaseous fuels, whose energy content is converted directly into heat or motion.
- Most waste plastics can be reprocessed at relatively low temperatures and energies…. but their relative low cost means their collection, separation, and recycling is often not cost-competitive with the production of new, virgin plastic. This is a critical point that will be explored in detail in sections related to recycling.
- Different plastics are considered by the public as being very similar, and their differences are often hard to distinguish by

eye… yet their tailored chemistries, mechanical properties, and additives content often tie each plastic to a specific use (which also makes them harder to recycle and reuse in multiple applications).

Thus, even though plastic products are sophisticated in their design and fabricated for even the simplest uses… most are still typically destined to become (very durable) trash.

At least one of these contradictions is directly addressed in this book: the use of valuable fossil fuels to create low-cost plastics. This is generally the case, though slowly in recent years, the existence of plastics is becoming less dependent on supplies of fossil fuels. More plastics are being made from renewable, biological sources. The growth rate of bio-based plastics is projected to be 20% through 2022, rising from an annual production volume of 2.05MM tons in 2017 to 2.44MM tons in 2022 [1]. Bio-based polymers right now include the increasingly popular corn ethanol-based polylactic acid (PLA). But the bio-based field may soon become dominated by conventional, well-understood polymers, such as polyethylene and polypropylene, that are based on feedstock made from plant resources such as Brazilian sugarcane instead of fossil fuels. For example, in 2010 Braskem SA opened a plant in Brazil to make 200 million lb/yr of sugarcane-based polyethylene [2]. Although these materials will initially be priced at a premium compared with fossil-fuel-based PE and PP, they stand as signs that the industry is indeed able to develop bio-based options for plastics. As will be discussed in later chapters, many biological resources are being considered for creating bio-based plastics, ranging from food crops to bacteria, and from agricultural wastes to algae. As we will see, key questions about these bioresins concern their properties and prices relative to those of fossil-fuel-based polymers — with many environmental, design, and social factors influencing the choices of manufacturers and consumers regarding bioversus non-bio-based plastics.

1.2   Plastics and the Consumer Lifestyle

Another plastics contradiction concerns consumers both in developed countries and, increasingly, in developing countries: Consumers have become incredibly dependent on plastic materials, even without understanding many of the details of their composition or how they are manufactured.

There is something almost fantastic about plastic products — their shapes and colors, their textures and smooth forms. In the more exciting early days of plastics, people might have even been tempted to reference science fiction author Arthur C. Clarke's "third law" from 1962: "Any sufficiently advanced technology is indistinguishable from magic" [3]. Plastics are no longer new; perhaps unfairly to scientists, they don't inspire awe. The most common popular polymers have outlived their inventors, and any sense of plastics' magic or fascination for consumers is lost behind the banality of the many products they are used in. Rather, a more common question overheard today would be: "Do we really need plastics?" or "Why do we need so much plastic?"

Thus, ignorance, indifference, hostility, or ambivalence might be better terms for describing consumers' views of plastics today. These views have resulted in the easy demonization of certain forms of plastics, with a complete obliviousness about other potential problems associated with the ways we use all materials in industrialized society — metals, wood, and minerals, as well as plastics. Humans have consumed more of all these materials in the past 50 years than in all previous years combined — at increasing rates. And the vast majority of industrial materials are not based on naturally renewing resources, unlike in 1900, when over 40% of all materials used in the United States was based on forestry, agriculture, or other renewable industries [4]. These trends, coupled with incredible population growth, have definitely stressed the environment.

Plastics simply reflect these trends, though perhaps more prominently because they are used for so many visible, high-volume consumer items. In fact, the average consumer is the plastics industry's biggest supporter, whether he or she is aware of it or not. The growth rates for plastics, both raw materials and end-products, continue to climb in all countries. As incomes rise, so does the use of plastics. People migrate to cities where increasing urbanization leads to requirements for more convenience. Supermarkets and other stores sprout up. In countries where populations are aging, there is increased demand for medical- and health-related products, many of which are made of, or packaged in, different types of plastics. Today, China is the leading producer of plastic with almost 30% of global market share, yet plastic waste generation is highest in developed western countries, the US in particular. There is no doubt that worldwide consumer interest in and dependence on plastic-based products will continue for as long as it takes to develop equally versatile materials of another kind.

Consider the ubiquity of plastics by looking at a day in the life of an adult in an industrialized country. When she wakes up, the first thing she touches is probably plastic: an alarm clock. Walking across the (nylon) carpet to the bathroom, brushing her teeth, washing her hair, and using other personal care items, she encounters a variety of plastics. The clothing she dresses herself in likely contains some polyester or another synthetic fiber. Her juice bottle and cup are plastic, as is her coffeemaker. She wraps her lunch in a plastic film or bag and places it in a reusable polymer-fiber sack. She drives to work in a car whose interior is almost completely covered in various polymer-based materials. She uses a coded plastic ID card to get into her workplace, grabs a (PET) bottle of water from the office refrigerator, and sits in front of a plastic-enclosed computer to spend the day tapping on plastic keys. Later, her evening at home or out on the town consists of similar contact points with plastics. The ubiquity of plastics was explored in detail in the 2011 book, *Plastic: A Toxic Love Story* by Susan Freinkel.

When looked at this way, the prevalence of plastics can be striking, though when we consider world plastics production growth since 1950, perhaps we should not be surprised (see Figure 1.1 below). In fact, their importance seems directly proportional to the degree to which the consumer does *not* notice that he or she is actually using a polymer-based product (especially in the case of beverage bottles, toothbrushes, shoes, cell phones, food packaging, and car interiors). Even though today most plastic items are banal and uninteresting to the user/consumer, a day without their use would be unthinkable. Some people have tried to demonstrate that we can live our lives without plastics, but they have met only limited success, facing continual frustration from not ever being able to totally succeed. With increased use, comes increased waste, yet as Figure 1.2 illustrates, we still manage to waste other materials more, including food.

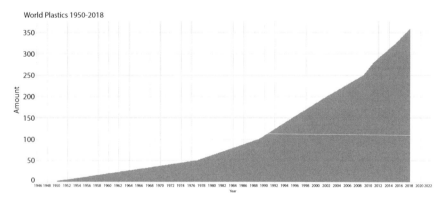

Figure 1.1 World plastics production (Source: Plastics Europe; World Economic Forum).

Plastics as a Contributor to Materials Waste Generated in the US, 1960–2015

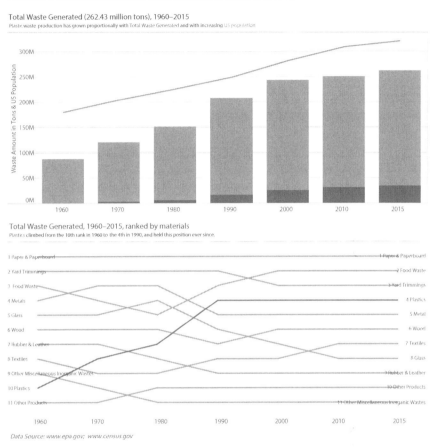

Total Waste Generated (262.43 million tons), 1960–2015

Plastic waste production has grown proportionally with Total Waste Generated and with increasing US population

Total Waste Generated, 1960–2015, ranked by materials

Plastics climbed from the 10th rank in 1960 to the 4th in 1990, and held this position ever since.

Data Source: www.epa.gov; www.census.gov

Figure 1.2 Plastic and waste trends, 1960 – 2015 (Source: US EPA).

1.3   Plastics Controversies

Despite the consumer marketplace's apparent desire for products made from plastics, the materials have faced a history of intense public scrutiny. Some early commercial plastics formulations were dependent on ingredients, such as heavy metals, with negative human health or environmental effects. Most of these compositions have since been eliminated, but enough similar applications remain to keep public watchdogs skeptical about the health impacts of plastics. The scrutiny continues and will continue, given plastic products' new compositions, new forms, and complete integration with everyday life. Acknowledging that this skepticism will continue is important for those who specialize in introducing new forms of plastics into the marketplace (even when these materials are based on natural, renewable resources).

Examples of plastics under scrutiny go back decades, and many controversies continue. Plastic additives, such as lead- and cadmium-based stabilizers and colorants, were once heavily used in polyvinyl chloride (PVC) and other plastics. Since then, there has been general acknowledgment that there is potential harm in using heavy metal-based additives and coatings, resulting in a strong trend in the industry to eliminate them. For similar reasons, electronics manufacturers are avoiding plastics that include halogenated flame-retardant additives, given concerns about the chemicals' environmental effects and persistence. And calls continue for eliminating the use of PVC itself, given its chlorine content and chlorinated-compound emissions during its manufacturing.

Another obvious example concerns many consumers' dislike of the use of expanded foam polystyrene products. The material is associated with possibly carcinogenic styrenic petrochemicals, ozone-layer-depleting blowing agents (once in common use), and its single-use, disposable, non-biodegradable nature. The feelings are intense enough that even the use of a foam coffee cup can draw someone's judgmental stare. A future alternative to achieving the lightweight insulation properties of expanded PS may be foamed PET or PP or even bio-based plastic foams. Entrepreneurs have entered the fray, offering alternative materials that promise to deliver the same benefits as PS. Vericool is one such company. Based in California, the company offers compostable materials made from extruded milled sorghum that are used in the refrigerated transportation sector. The company has attracted interest and investment from larger entities focused on transporting items within 24-96 hours where temperature management is critical [5]. Although PS can be recycled successfully, notably by companies like Agylix in Oregon, it is losing market share both from outright bans and from brands switching to other polymers such as PET and PP.

Three other recent controversies related to plastics are of interest. These deserve more discussion below, and have been discussed frequently in the news media over the last few years. Unless they and some lesser-known issues can be resolved, these and other issues will likely keep plastics in general under intense public/governmental scrutiny perhaps even transferring over to polymers made from renewable, bio-friendlier resources or processes.

1.3.1 PVC and Phthalate Plasticizers

Polyvinyl chloride, or simply PVC or "vinyl," has progressed down a rocky road from its original high mountaintop location of being one of the most affordable, adaptable, and widespread polymers for plastic products. Since then, both the material's origin and end-of-life have been questioned in terms of environmental and health impacts. When concerns about persistent industrial chemicals in the environment became more prominent over the past few decades, the potential toxicity of PVC's chlorine chemistry itself was scrutinized. Its production depends on using cancer-causing vinyl chloride, which can escape into the environment if not carefully controlled, along with small amounts of chlorinated dioxins (which are carcinogenic and persistent in the natural environment). Heavy metal-based additives (mentioned above), once commonly used in vinyl, are largely being phased out, but not after somewhat damaging PVC's reputation. And vinyl products are not of much interest in major recycling efforts, given all the varieties of PVC compounds in use and the challenges involving their collection and separation.

Yet no final resolution about vinyl's use is evident. It continues to be used in high volumes for siding and roofing in building and construction, and in some consumer products and packaging. However, governmental and environmental group efforts are having some effect in discouraging its use in electronics and consumer goods. In 2010, the European Union moved to require special assessments of PVC's use in electrical devices, outside of the EU's already restrictive REACH (Registration, Evaluation and Authorization of Chemicals) system and RoHS (Restriction of Hazardous Substances) directive [6]. This news is nearly the only kind of media attention vinyl gets today, and millions of consumers by now have a negative PVC taste in their mouth from the publicity.

In the 1990s, health concerns arose related to the phthalate chemical plasticizing additives used to soften PVC. Commonly used phthalate plasticizers are not chemically tied to the polymer backbone, so they have a tendency to migrate to the surface of the product. Their effect on children was of particular concern, and children were the driving force of the controversy, because they tend to place flexible vinyl toys and products into their mouths. After years of studies about how phthalates disrupt the endocrine system of humans, and counter studies saying the opposite (often funded by an industry that relies on using phthalates for PVC), regulations began to be passed that restricted the use of common phthalates. Apart from outright bans in some western countries, in 2009 the US Consumer Product Safety Commission (CPSC) restricted to very low concentrations a number of common phthalates in products used by children under twelve years old. In 2017, however, this same body issued a rule "prohibiting children's toys and child care articles containing more than 0.1 percent of certain phthalate chemicals" [7]. This brought to eight the total number of phthalates that are restricted from use in children's toys. In the European Union, RoHS is adding four phthalates to their list of restrictive substances [8]. In response, additive suppliers have created lines of plasticizer alternatives, some of which are bio-based or which bond to the PVC polymer matrix [9]. Perhaps there are lessons to be learned from the history of PVC as new bio-based plastics are introduced. Some resin suppliers are already considering questions like the following before the large-scale marketing of their fossil-fuel-based resins or their bioresins:

- What compounds are emitted during the production of the raw polymers, and during their compounding? Are they dangerous to workers? Would any regulatory agency consider them so?
- What additives are compounded into the plastic? Are any controversial? Have they been tested and confirmed to be non-toxic in every situation in which a user might encounter them? Are they tightly bound in the final plastic composition, or can they be extracted from the plastic part during use?
- Can a recycling infrastructure potentially be created for the material, if it does not already exist (or a composting infrastructure, if the material is compostable)?

1.3.2   Plastic Shopping Bags

The issue of banning single-use plastic shopping bags illustrates the concept that the more an inexpensive plastics technology grows in popularity, the greater the potential for controversy. There is a reasonable motivation driving the desire to reduce or eliminate the use of lightweight plastic shopping bags: reducing litter. Polyethylene blown-film bags are lightweight and designed mainly for a single use, thus they are treated as low value, trivial products by the consumer. They are also easily littered, intentionally or accidentally, and are easily transported by the wind and waterways far from where they were littered. They are also resistant to degradation by weather or water. Plastic litter in general is colorful, durable, noticeable litter, especially when it is reported as collecting on beaches or in ocean gyres. These problems are argued very persuasively and visually, with pictures of wildlife harmed by ingesting plastic.

This explains the rash of plastic bag bans proposed or enacted by regional governments around the world (see Figure 1.3). By the start of 2019, hundreds of bans of conventional, non-biodegradable grocery bags had gone into effect in communities across the globe. In the US, the state of California has enacted an outright ban, which did not lead to the end of the world, as suggested by some members of the plastics industry [10]. Some governments collect a fee or levy a tax on each bag used. Corporate retailers have even developed their own self-imposed bag bans as shoppers at grocery stores around the US are now offered paper bags for free, or the option to purchase heavier-duty plastic bags for a fee.

Yet despite litter concerns and bag bans, the use of plastic packaging overall has continued to grow. There are other contradictions in the bag ban trend. One often cited observation is that by eliminating single-use plastic bags, bans encourage the use of more paper bags, which are heavier and less energy-efficient to manufacture. In a study concerning the banning of plastic carryout bags in Los Angeles County, researchers calculated that replacing 6 billion plastic bags per year with 4 billion paper bags would create additional CO_2 emissions equivalent to tens of thousands of

States with Enacted Plastic Bag Legislation

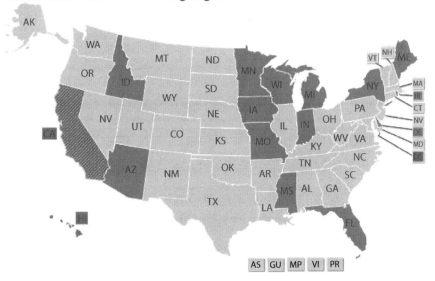

Figure 1.3 US States with enacted plastic bag legislation (Source: NCSL.com; as of Jan 17, 2019).

additional vehicles on the road. Moreover, paper bags that are landfilled rather than recycled quickly degrade, seemingly a good thing except for the production of methane, a potent greenhouse gas that many landfills are not equipped to capture [11]. There is no shortage of life-cycle analyses illustrating that plastic bags are more efficient and create less CO_2 emissions than paper counterparts.

Another opposing argument is that any fee levied per bag, such as $0.05 to $0.25, unfairly impacts low-income people as a regressive tax, while not itself accumulating into a significant source of government revenue [12]. As bag bans began to take effect, several sources trumpeted their success or failure, with the plastics industry, through various lobbying efforts, stressing their preference for job protection while creating more uniform standards for waste management and recycling [13]. Ban proponents point out the declining use of plastic bags as evidence that bans work in changing consumer behavior to the benefit of the environment: fewer plastic bags, fewer such bags littered.

There are also counterproposals for fixing the problems related to plastic bags. Industry organizations such as the American Chemistry Council, the American Progressive Bag Alliance, and PLASTICS (the primary plastics industry association) argue that strong public education programs that discourage bag littering and promote bag recycling are more effective approaches to handling the issue. In response to future bag bans in Europe and elsewhere, bioplastics producers have made investments in capacity, preparing for an increase in demand for blown film made from bioresins. Bio-based and/or biodegradable plastics may be sustainable alternatives for shopping bags, whose manufacture uses tens of millions of oil-equivalent barrels of fossil fuel each year. This sector has seen tremendous growth over the past decade and is set to become worth $1.7bn by 2023 [14]. (By comparison, polyethylene alone is a $164bn market) [15]. Yet bioplastic bags still cost more than the PE bags and use more water and energy to manufacture. Many biodegradable bioresin bags do not completely or quickly degrade when littered, scarcely helping to solve the littering problem. Meanwhile, more consumers are getting into the habit of using reusable bags, some made of cloth, some made of heavier-duty plastic. With the advent of COVID-19 and renewed interest in and appreciation for disposable plastics, there have been numerous reports and studies on the topic of reusable bags, suggesting/proving/refuting that these bags can harbor harmful bacteria, including *E. coli* [16].

This continuing plastic shopping bag saga indicates at least three questions to consider when making sustainable plastic product decisions for single-use products in particular:

- Is the product designed so as to seem very disposable and trivial, making a littering controversy inevitable? Or can its recyclability and multiple uses be emphasized by its design or marketing?
- If the product is likely to become litter, is its material formulated to degrade under non-composting, natural conditions? If not, what will its effect on wildlife and the natural environment likely be if littered?
- Can a significant percentage of post-consumer recycled material be used in the single-use product, emphasizing that there is value in and a need for a stream of recycled materials of the same kind?

1.3.3   Health Effects of BPA (Bisphenol-A)

A distant relative of the PVC/phthalate controversy is the concern raised about residual bisphenol-A in polycarbonate (PC) containers and epoxy-based metal can liners. BPA is one of the building blocks of PC and epoxy polymers, and residual BPA can be detected in measurable amounts in final products and on their surfaces. Like phthalates, BPA has been classified as an endocrine disruptor. Various laboratory studies have pointed to BPA as disrupting the glandular systems and development of mammals in various degrees, depending on concentration and intake amounts. Studies continue to emphasize widespread exposure of the general population to BPA and the possible effects on children, though the studies are heavily scrutinized and criticized by industry groups. In February 2018, the National Toxicology Program, a division of the FDA, released its findings following a two-year study on the subject. In short, the preliminary findings supported the FDA's determination that current use of BPA in consumer products was safe [17]. The statement was challenged by several groups, including the Endocrine Society and researchers at various university labs. Yet the focus on BPA in plastic is perhaps too narrow. According to John Warner, President of the Warner Babcock Institute for Green Chemistry, "There's more BPA in a single thermal paper receipt than the total amount that would leach out from a polycarbonate water bottle used for many years."

Still, the point may be entirely moot. Since at least 2008, manufacturers and retailers began phasing out PC baby bottles. As of 2018, dozens of US states and EU member countries had established laws restricting BPA in various types of containers. Reports about BPA studies have led to various bans of PC products in various western countries and US states and cities. The bans focus particularly on containers used by children, who are in sensitive phases of endocrine system and hormonal development. Various large retailers are also phasing out all water- or food-related PC products, responding quickly to mass media reports, public concerns, and governmental warnings. Canned foods appearing with the label "BPA free" are appearing on grocery store shelves. Even the common 5-gallon PC water cooler bottle used in offices around the world is being redesigned to be molded in non-BPA-containing copolyester resin.

Thus, public opinion and even the views of some of its own member companies have turned against the plastics industry's official positions on BPA. For the plastics industry, the issue has become a "damage limitation exercise," given all of the common plastic products BPA can be found in, the vast majority of which are not food/drink related [18].

The process by which BPA-based plastics have become undesirable includes key themes common in chemical controversies. One basic issue is that BPA is a man-made chemical, widely used by the often distrusted chemical industry. BPA is also somewhat mysterious — it may be present in our food and drink containers, but it cannot be detected by our senses, and consumers do not know when they are being exposed to excessive amounts of it. Inside the human body, its effects sound disturbing; studies indicate it can mimic estrogen, a hormone whose proper levels are critical for the sexual development of fetuses and children. This makes the issue even more compelling and emotionally driven.

Given these conflicting issues, biases, and emotions, a couple of questions might be addressed before any new plastic or product line is introduced for a human-contact application:

- In addition to its additives, what other residual materials in the plastic might have questionable health-affecting or environmentally controversial backgrounds, inviting scrutiny?
- Can issues with the plastic's composition be connected with recent plastics controversies? And if so, will that hinder its acceptance by retailers or consumers?
- Should manufacturers use the precautionary principle when faced with materials that have questionable health effects? How will they justify investments in alternatives that could require long validation and approval timelines?

Negative reactions from all controversies about plastics or their ingredients ultimately spread and become intensified in the general public discourse. Some of the mainstream media reports are written by journalists who tend to oversimplify and overgeneralize, lumping more common plastics in with the controversial types, and even inaccurately implying that all common plastics share the same questionable characteristics. The writers sometimes even imply that a number of common modern health ailments may be the result of plastics use alone, despite all the other materials, pollution, technologies, radiation, stress, and processed foods that people are exposed to. Such writers and their hasty generalizations can be taken to task in the industry press but even with the Internet, such technical corrections [19] are not likely to be understood (or even read) by the general public audience of the original, more dramatic mainstream reports. It should be pointed out, however, that given increased awareness and knowledge of plastics, current coverage tends to be more balanced than it once was.

Current controversies also invite increased future scrutiny on plastics when other health issues arise. Sometimes intense scrutiny is helpful, contributing to safer products in the long term. But any questionable plastic compound ingredient can quickly become the focus of intense studies funded by various groups. The intensity of the scrutiny may border on irrational reasoning, as in a reported case of excessive antimony detected in packaged fruit drinks in Denmark [20]. In this case, the main focus was the antimony trioxide used as a catalyst in the production of PET packaging — making the PET the prime suspect source — even though high antimony was also found in *non-plastic* packaged fruit drinks.

The mainstream, traditional plastics industry is typically the focus of this kind of scrutiny. But although manufacturers of bio-based plastics may get better press because of the natural origin of their materials, even they should anticipate greater scrutiny once their products become more established.

1.4   The Desire to be Green

Given these kinds of controversies, plus consumers' other environmental concerns about plastics, we again notice contradictions and conflicts. Consumers want to be green, but they also apparently want plastic products. This section (and the remainder of this book) looks at how these often conflicting intentions might be resolved.

1.4.1   Consumer Interest in Sustainability

Consumers are interested in environmental sustainability, though their behaviors may not reflect their interest. In one recent survey, over 80% of consumers expressed concern about the environmental impact of their choices, and said they thought retailers' environmental efforts are important.

But fewer than 10% would sacrifice convenience to buy sustainable products. And for many, the term "sustainable" has become part of the daily lexicon, diluting its meaning and impact. Most people feel recyclability is important, though in general consumers do not display diligence in their green behaviors overall, especially behaviors related to plastics. For example, consider…

- their rampant littering, which clutters the sides of roads, clogs waterways, and mars natural scenery (and here plastics make up a minor but very noticeable proportion of littered materials);
- their inclination to throw easily recycled bottles into garbage cans rather than recycling bins;
- their favoring of factory-made products that are heavily packaged with plastics and other materials,
- their constant replacement and disposal of electronic devices and media, most of which are encased in plastics, and
- their addiction to convenience, fed by the ease and simplicity of e-commerce,

to cite just a few observations. These real-world tendencies might not change much, and plastics producers and designers interested in sustainability must design around them as much as possible to "pollution proof" devices and packaging, and to make their recycling as easy as possible.

Recycling is something every school child learns about at an early age, but our society recycles plastics relatively poorly, especially in the United States. For example, despite some gains in recent years, the United States' recycling rate for PET and other bottles has remained under 30% through 2020 (see Figure 1.4 below). Europe does better, especially in PET

Figure 1.4 2017 Recycling rates for PCR PET Containers (Source: Plastics News, American Chemistry Council).

bottle recycling, with a rate of about 58% [21]. Generally speaking, the EU approach has favored a combination of regulations and incentives. The US approach has been hobbled by a patchwork of overlapping and conflicting efforts from the municipal level to the state level. Despite evidence that suggests container deposit laws improve recycling rates many ballot initiatives across the US have failed, mostly due to heavy lobbying from beverage companies and their associations [22]. But experts say a sub-50% PET recycling rate is still not high enough to offset the net carbon emissions of processing, collecting, and reprocessing PET — or to bring the recycled PET's price low enough to be really competitive with virgin resin. Only collection rates of over 50% will result in a noticeable amount of material being reused in new PET products, including not just fibers and low-end products, but also new bottles and food packaging. As major brands continue to make public commitments to increase the use of recycled materials, the market has failed to react, leading to a supply shortage which in turns drives up the price of recycled resin. This has led to vigorous debate in policy creation, as industry and state legislatures wrestle with language that will mandate additional recycling without placing job-threatening burdens on producers.

On the other hand, given their desire to be environmentally sustainable, efforts are being made to lead consumers to greener products and companies. Going back to 2007, major US retailers and suppliers of consumer products, starting with Walmart, introduced sustainability "scorecards" for their suppliers. These systems evaluated the carbon emission and resource and energy "footprints" of their products and operations. Many companies followed suit, the result being an increase in data and information, not to mention burdensome bureaucratic requirements for many small- and medium-sized businesses. Meanwhile, marketers are responding to consumer interest by touting environmental claims for products and processes (though some of the claims are better supported than others).

These seemingly positive developments can be read in complex ways, however. Demands for environmental sustainability are often motivated mainly by a company's simple interest in reducing energy, shipping, or material costs. Or green efforts may be used to increase sales by stressing relatively minor sustainability improvements using major marketing campaigns. Some claims fall into the unsubstantiated "greenwashing" category of exaggeration, pseudo logic, or lack of objective verification.

Greenwashing has become such a concern that the US Federal Trade Commission proposed "Green Guides" guidelines in 2010 about what specific green marketing language is required by companies which exhort the environmental values of their products. The term "eco-friendly" now must be more substantiated and made specific. For example, the term *biodegradable* by itself is too vague to be clear unless qualified by testing, though it has been wielded as a magic word that automatically attracts concerned consumers. There have been several legal judgements in California and Germany against companies (including Walmart) [23] who made unsubstantiated or incorrect claims about their materials and additives.

More recently, companies and individuals in the plastics industry have become highly-attuned to a new evolution of greenwashing — virtue signaling. This term refers to public pronouncements of environmentalism, usually in the form of removing or reducing plastic while ignoring the very real GHG-related impacts of those decisions. While it is undeniable that plastics pollution (otherwise known as littering) is real, has captured the public imagination, and has helped to create consumer-led campaigns against certain products (shopping bags, plastic straws, stirrers, etc.), our soundbite society is not patient enough to understand the complex nuances of this phenomenon. Images of marine life choking on plastics evoke strong, emotional responses, yet the ocean is littered with sunken ships and a variety of human waste. Many individuals and companies are taking real steps to address the littering of plastics on land and sea, yet the source of the problem is individual behavior. According to Keep America Beautiful, about 85% of littering is the result of individual attitudes [24].

1.4.2   Sustainability: Views and Counterviews

Generally, this interest in being green makes sense. It can be assumed that everyone, theoretically at least, wants environmental sustainability. People want a clean, habitable earth with healthy places to live in, plus healthy wildlife in natural areas and oceans. However, the specific issues that determine or affect sustainability do not have consensus. The concept that the earth is warming because of excess greenhouse gas emissions is mostly accepted by people around the world. However, some questions exist about the extent to which the warming is happening, or on whether human activity is mainly causing it and can stop it (with some people perhaps having an innate belief that human beings could not possibly have such influence over nature). Discussions are also related to the effects of the warming, with some people perhaps even doubting that the warming will be a bad thing overall.

With plastics and other consumer materials, opinions (informed and otherwise) range from the simple to the complex. As we have seen with the controversies described above, educated people, though they use plastics every day, commonly have a fundamental notion that plastics are bad for environmental sustainability, except when plastic products are made from natural materials or recycled plastics. They may believe recycling is always good, as they learned in school. The emotions behind these beliefs range from zeal (or hostility) to indifference. In the general public, the interest in the deeper science behind these issues is limited, even when the news media draws attention to controversial studies about plastics. Concerns like the BPA issue explained above have led to the media's tendency to report on every obscure negative plastic study, with most of this reporting being oversimplified, confusing consumers about what to do.

For the deeper analysis that is required, sustainability and plastics issues have drawn many different approaches from various researchers and institutions. Some of this research is politically driven by environmental groups, free-market theorists, and the plastics industry itself — each with its own inevitable bias — though this bias does not automatically make their arguments weak or illegitimate. University-based research is invaluable and is intended to be less biased, digging down into the deepest layers of specificity about polymer science and the environment. In 2016, the American Chemistry Council commissioned a study [25] from Trucost, a consultancy specializing in environmental, social, and governmental (ESG) data analysis. The results showed that, if plastic were replaced by alternative materials such as glass, aluminum and paper, the carbon emissions would be four times greater. Critics, however, were quick to point out that not all plastics are created equal, and that simply by creating a large, composite view of "plastics," the report glossed over other thorny problems. One of these, recycling and its economics, will be covered in more detail later in the book.

However, mainly in the public discourse, traditional fossil fuel plastics receive much less good press than newer plastics based on renewable natural materials. People outside and inside the industry are excited or at least curious about bioresins and whether they could solve the environmental negatives associated with traditional plastics. There are also genuine contrarian arguments about the use of chemicals and plastics in general — and about bio-based materials in particular. Below, a few of these arguments are summarized and addressed, revealing the assumptions behind the context of this book.

What Global Warming? After many years during which the concept of global climate change has been discussed, debated, and studied, there has been at times the sense that consensus has been achieved: that it is clear that greenhouse emissions caused by the burning of fossil fuels for over two hundred years indeed is causing an overall warming of the Earth's climate. In 2009, the US Environmental Protection Agency (EPA) cites the Intergovernmental Panel on Climate Change (IPCC) as stating the warming of the climate is unequivocal, and the EPA has taken the position that the warming is caused by greenhouse gas emissions, which can potentially be controlled and reduced. As a government office, however, the EPA has been at the center of the "climate wars" with different administrations strengthening or weakening the mandate of the agency. The IPCC position has not changed over the past decade, and the US government's "Fourth National Climate Assessment" released in 2018, reinforces risks posed by climate change.

Yet skeptics and minority views persist, and many are not simply politically motivated. A small number of reasonable and informed researchers do continue to question the accuracy of climate models and data used to support the idea that human activity is causing significant warming. The climate is more complicated than the models predict, some argue, thus overstating the amount of warming that is happening. Anyone who has read *The Black Swan* by Nassim Nicholas Taleb will recall his skepticism of models, especially when the error rate is not acknowledged. Generally speaking, in Taleb's view, uncertainty itself "radically underscores the case [for precaution] and may even constitute it" [26]. These counter-arguments might mean that major reductions in the use of fossil fuels are not critically important; and likewise, that developing alternative, renewable sources of materials and energy may be a more expensive exercise than can be justified.

This book is not meant to verify and/or support either those who agree with or disagree with various global warming projections. To some extent, the issue of using fossil fuel-based or bio-based plastics mostly sidesteps the issue, since plastics production is a smaller source of greenhouse gases than other sectors of the economy, including transportation, general electricity production, and heating. The book does, however, take seriously the possible disastrous consequences of global warming, if the worst scenarios play out. And it takes seriously the interests and views of consumers and industries that are interested in reducing the role of fossil fuels and the greenhouse gas impact of plastics use and production. Just as importantly, this book looks to a future when alternatives to fossil fuels will be necessary — when fossil feedstock sources begin to become so depleted and expensive that renewable resource feedstocks become cost-effective options for producing plastics.

Aren't we running out of oil? Conventional plastics production is dependent on access to feedstocks produced from steady supplies of oil and natural gas. These resources are technically finite on Earth, and humans have not found a competitive option to using these fuels for heating, transportation, and chemicals.

Yet, although fossil fuel extraction and production has become more technically challenging, fossil fuels continue to be found in various locations and extracted in various ways, such as from shale. Crude oil's price and the price of natural gas have continued to be stable enough to support current plastic demand, keeping plastic products affordable. In the US specifically, the price of feedstocks for polymers has decreased dramatically thanks to the availability of cheap ethane which is derived from natural gas. Over $180bn has been invested globally in new petrochemical cracker plants with the US states of Texas and Louisiana set to see industry investments of almost $50bn along the Gulf Coast [27]. This shift has created an enormous advantage for US-based plastic resin suppliers who now account for the majority of exports, leading to an overall trade surplus in the plastics industry.

People and jobs are more important than the environment, aren't they? It is still common to hear the argument that jobs for people are more important than environmental sustainability. We hear that costs or taxes must be cut in order to create jobs, or to prevent additional job losses. Government subsidies for sustainability are frowned on if they do not create jobs immediately. This focus tends to hurt efforts at sustainability. For instance, efforts may be abandoned for increasing the use of bio-based materials if they initially invite extra costs that make them uncompetitive with traditional materials, even in a $100-per-barrel oil economy.

Interest in sustainable materials is growing, however, and market opportunities exist. Efforts in environmental sustainability can sometimes become new ways of saving money and better ways of using materials, both fossil fuel- and bio-based. This book assumes that this interest in sustainability will continue in future years and is likely to grow among people within the plastics industry and plastics using sectors. This interest will translate into new markets which will support bioplastics use, as existing markets for conventional plastics remain strong or grow themselves.

Is a sustainable use of resources even possible? Researchers who take the long view argue that western (and now eastern) societies have increased their use of material and energy resources over time, and nothing will likely stop that trend. In terms of energy use, a strange relationship known as the Jevons Paradox seems to exist: the more improved the energy efficiency of our technologies is, the more the total demand for, and use of, energy is. For example, improvements in automotive engine efficiency since the 1970s have been accompanied by a large increase in total miles driven each year, and thus increased gasoline consumption and greenhouse gas emissions. If this relationship is true, then steps toward sustainability may simply encourage people and the economy to demand and use more and more materials and energy. Ironically, the increasing use of plastics for light-weighting in the transportation and building sectors is leading to concern among members of the petrochemical industry as plastics will account for a greater share of oil and gas feedstocks. A new type of plastics paradox, perhaps [28].

An assumption behind this book is that materials use, especially plastic products use, will continue to increase globally — whether more efficient, sustainable approaches are employed in manufacturing them or not. Growing economies, such as China and India, will increase their demand for plastics, perhaps to extreme levels. At the same time, no matter what raw materials they are based on, plastics will also continue to be improved in terms of their performance in applications. Other improvements will relate to the amount of energy and resources they consume when produced, how easily they can be recycled, and perhaps how well they can be composted instead of landfilled. Thus, energy and resource conservation is an adequate reason for pursuing the development of biobased plastics and greater recycling. (And it can only be a good hedged bet for the plastics industry to possess technologies that give it some independence from fluctuating oil- and natural gas-based feedstock costs. See Figure 1.5 for 20-year commodity price changes.)

Recycling always makes sense, doesn't it? Despite the average person's belief that recycling is always a positive for the environment, there are major hurdles involved in collecting and recovering recyclable plastics from the waste stream. Plastics are recovered at lower rates from US municipal solid waste than all other major material types, even with the many curbside recycling programs now in place [29]. Separating uncontaminated plastics by type and form is not easy for consumers or recycling facilities. Also, there are often technical limits on the amount of recycled resin that can be used in a given product, and the range of products containing recycled content is limited by the quality of available recovered material. The costs and difficulties of plastics recycling — and the relatively low price of conventional virgin resins — can mean that sometimes recycled resin costs as much as virgin plastic resin, despite the strong demand for recoverable material by recycling operations around the world. As mentioned above, increased commitments from major brands to use more recycled plastics should be the signal the market needs to stimulate investment in new capacity. In August 2019, the National Association for PET Container Resources (NAPCOR) bluntly stated that supply will not keep up with demand without a sea change in recycling infrastructure, starting with the

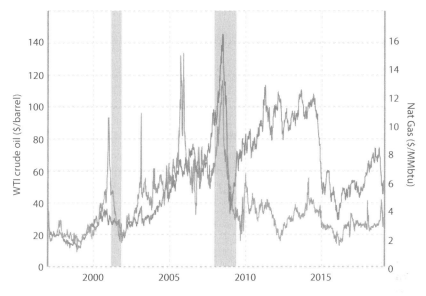

Figure 1.5 Crude oil vs natural gas: 10 year prices (Source:www.macrotrends.net/2500/crude-oil-vs-natural-gas-chart).

collection systems. It is becoming apparent that the traditional curbside recycling model is not sufficient to enable a steady supply of quality, clean, sorted materials. US states (and EU countries) with container deposit schemes consistently show higher recycling rates, though complexity rears its head here, too, as municipalities could lose funding for recycling if valuable materials are removed from their collection stream. Plastic, glass, and aluminum containers are not viewed with equal value by all, yet some expansion of deposit laws seems likely despite well-funded opposition.

But if efficiently done, plastics recycling can be a valuable way of reusing the building blocks of polymers and their chemical bond energy, rather than wasting this resource by burying it in a landfill. Additional studies confirm that recycling has a notable impact on avoidance of GHG emissions [30].

The recycling potential of both conventional and bio-based plastics is strong but mostly underdeveloped in most countries. There is much room for improvement. Meanwhile, recycling companies in China, the United States, and Europe are seeking more recycled plastics to fill their extra capacity. Public education by the plastics industry about recycling can be improved, creating a greater stream of higher quality material. The public is, however, perhaps skeptical of industry-led programs given high-profile lawsuits and industrial accidents where chemicals, if not plastics, were involved. Programs such as Recycle Across America are leveraging Hollywood star power to draw attention to simple changes such as correct labeling for bins. And a strong recycling industry can supply material that serves to hedge against spikes in raw material prices.

If bioplastics cost more, and we're not even sure they're truly sustainable, why should we use them? Current common bioplastics are relatively expensive because of the fuel, fertilizer, water, and energy required to grow their raw feedstock material (plants). To make them more price competitive with fossil fuel-based plastics, subsidies may be needed. Critics have argued that government-subsidized materials are in fact economically *unsustainable*, distorting free market practices that normally reward materials and processes having the lowest costs. The fact that producing bioplastics requires many resources may also mean that they are environmentally unsustainable as well [31]. Various studies have tried to show that bioresins do have a lower environmental impact. However, the complex arguments required for making good life-cycle evaluations have made these studies vulnerable to criticism and rebuttal (see Chapters 2 and 3).

Admittedly, it is possible that the newest generations of bioplastics in production do not have strong economic or environmental sustainability, at least not with current technologies and volume demand. The number of factors involved in evaluating the true environmental life-cycle impacts of bio- and fossil-based plastics is staggering. However, there is a driving force that may help bioresin production become more efficient, more competitive, larger in scale, and greener: the consumer. The environmental ethics of some consumers indicate that at least for some products in some markets, there is strong market demand for bioplastics. Some consumers simply are interested in buying plastic products that are based on plants or other natural resources, even if they cost more or cannot be proven to be unequivocally environmentally sustainable. Only time will tell if makers of bioplastics and their applications will find durable long-term markets and lower costs as a result of this consumer interest.

1.5   The Course of This Book

As the following chapters will discuss in more detail, the criteria for determining what makes a plastic product "green" can include many measures, including factors related to its source feedstock, the carbon dioxide emissions in its manufacturing, its recyclability, its biodegradability/compostability, and the toxicity of its composition. All "cradle-to-grave" lifecycle impacts must be considered: the origins of the compounded polymer material and its processing into a useful product; the product's environmental impacts during its use-life (from the time it comes out of its manufacturing facility to when it is shipped to a waste or recycling facility); and the environmental impacts from the product's disposal/recycling phase.

This book's structure attempts to serve as a template for a thorough, thoughtful analysis of plastics and sustainability. It covers the lifecycle factors influencing the selection of various plastic materials for minimal environmental impacts, with special emphasis on comparing the impacts of fossil fuel-based plastics with new biologically sourced polymers, fillers, and additives. Each aspect determining the environmental impacts of plastics (and the chapter which focuses on it) is summarized below:

The Life Cycles of Plastics (Chapter 2): All manufactured materials and products have certain costs and impacts associated with their creation, use, and disposal (or reuse). There are numerous ways of evaluating and measuring these impacts. These are based on analyzing all the material and energy flows that are used for producing and using the material or product, as well as the costs or value, if any, that the discarded product or material might have after it has been used. Chapter 2 will review principles for evaluating the fundamental impacts of the raw feedstocks used in polymers, and summarize ways in which the total lifecycle impacts of plastics have been measured through approaches such as lifecycle assessment (LCA). It will also discuss the limitations of LCA and the factors not easily accounted for when speaking about plastics' impacts, such as biodegradability and recycling.

Material Composition (Chapter 3): Various polymers require various feedstocks made from fossil fuels, or more often now, bio-based raw materials. These feedstock molecules are combined to form polymers, to which additives and fillers are added, creating plastics. Each kind of polymer/plastic impacts the environment slightly differently. This chapter will attempt to compare and contrast the impact and value of producing various commercial plastics and the polymers they are based on. It will give special attention to the composition and properties of bio-based polymers, relating their properties to those of more familiar fossil fuel-based polymers, to create a context in which they may be compared. Questions also will be addressed about the toxicity of the materials' components, and about how the various materials are appropriately durable for use in various applications.

Applications (Chapter 4): This chapter will discuss several real-life example applications in which sustainability was a factor in the choice of plastic material. The application sector areas touched on include packaging, automotive, construction, medical, electronics, and agriculture. The discussion will attempt to draw "lessons learned" from the examples that can be applied to new applications.

Design (Chapter 5): Product design decisions can reduce (or increase) the environmental impact of a plastic product. Subtle changes in a product's dimensions or in the way various materials are combined or assembled together can have large impacts on recycling and material usage. This chapter will provide basic guidelines for "green" plastic product design.

Material Selection (Chapter 6): This chapter will combine the elements of Chapters 2–5, showing how various factors — including environmental impacts — can be compared and weighed for different plastic application areas. Its aim is to assist in the process of determining the optimum choice of material in terms of performance, quality, cost-effectiveness, and, especially, sustainability — a factor often not emphasized in traditional material selection methodologies.

Processing Energy and Waste (Chapter 7): Significant energy can be consumed when creating plastic compounds and when forming the materials into products. Because bio-based resins and fossil-based resins use essentially the same conversion technologies, this chapter mainly looks at methods used for processing and reprocessing plastics in general. Available alternative technologies and practices are presented for efficiently recycling process scrap and post-consumer resin, and for reducing the energy and water costs of conventional conversion processes.

Conclusion (Chapter 8): The final chapter presents an overview of current worldwide trends and obstacles that relate to efforts in reducing the overall environmental footprint of plastics. It will propose basic conclusions about current trends, indicating directions for the future as well as suggesting goals the industry should consider targeting, and actions it should take to enhance the sustainability profile of plastics.

References

1. https://www.european-bioplastics.org/global-market-for-bioplastics-to-grow-by-20-percent/
2. Sherrard, A., Braskem to supply ethanol-derived bioplastic to LEGO Group's "Botanicals". *Bioenergy Int.*, 2018, March 9.
3. Jones, A.Z., 2011. 'What are Clarke's laws?' About.com. http://physics.about.com/od/physics101thebasics/f/ClarkesLaws.htm, 2011.

4. EPA, Sustainable Materials Management: The Road Ahead, 2009, June. United States Environmental Protection Agency.
5. www.vericoolpackaging.com
6. Nuthall, K., Europeans might target PVC in electronics, in: *Plastics News*, 2010, June 7.
7. CPSC.gov, October 20, 2017.
8. http://www.intertek.com/blog/2018-07-24-phthalates/
9. Martino Communications Inc., Teknor Apex now offers diverse range of alternatives to phthalate plasticizers for vinyl toys and child-care products, (Press release), 2009, June 22.
10. https://www.latimes.com/opinion/editorials/la-ed-plastic-bag-ban-anniversary-20171118-story.html
11. Greenhouse gas emissions will increase massively if Los Angeles County bans plastic bags and permits free paper bags, 2010, January 1. savetheplasticbag.com.
12. Higginbottom, J., (2010, May 7), Bag taxes disappointing in debut, in: *Fiscal Fact (no. 224)*, Tax Foundation, www.taxfoundation.org, 2010.
13. https://news.bloombergenvironment.com/environment-and-energy/plastics-industry-opposes-patchwork-of-rules-as-bag-bans-expand
14. IHS Markit news release, As Plastic Regulations and Bans Increase, Market Value for Biodegradable Polymers Exceeds $1 Billion and Will Rise Sharply by 2023, July 26, 2018.
15. Freedonia Group; Statista
16. https://www.solidwastemag.com/blog/covid-19-has-resurrected-single-use-plastics/. July, 2020.
17. https://www.fda.gov/NewsEvents/Newsroom/PressAnnouncements/ucm 598100.htm
18. Tolinski, M., Clear alternatives, transparent motives, in: *Plastic. Engineering*, 2009, March.
19. https://www.plasticstoday.com/extrusion-pipe-profile/fear-plastics-and-what-do-about-it/13536413858942/page/0/1
20. Smith, C., Fruit drink study finds antimony, but no cause, plasticsnews.com. 2010, February 25.
21. Johnson, J., NAPCOR: US lacks recycled PET to meet consumer brands' pledges, in: *Plastics News 2019*, 2019 August 29.
22. https://ballotpedia.org/massachusetts_expansion_of_bottle_deposits_initiative_question_2_. 2014.
23. https://www.environmentalleader.com/2017/02/greenwashing-costing-walmart-1-million/
24. https://phantomplastics.com/plastics-the-environment
25. Plastics & Sustainability: A Valuation of Environmental Benefits, Costs, and Opportunities for Continuous Improvement, Trucost, 2016.
26. Taleb, N., The Black Swan: The Impact of the Highly Improbable, 2007/2010. Random House.

27. www.analysis.petchem-update.com
28. Butler, N. and Tipaldo, E., Unlocking the Plastics Paradox. *Resour. Recycl.*, February 2019.
29. US EPA, Facts & Figures about Materials, Wastes and Recycling. www.epa.gov.
30. Turner, D., Williams, I., Kemp, S., Greenhouse gas emission factors for recycling of source-segregated waste materials. *Resour. Conserv. Recycl.*, 105, Part A, Dec. 2015, pp. 186-197, Elsevier.
31. Jones, R.F., An overview of environmental alternatives as viewed by a plastics industry economist, ANTEC 2009 (Proceedings). *Soc. Plast. Eng.*, 2009.

2

Plastic Life Cycles

With manufactured products, a cradle-to-grave life cycle assessment (LCA) requires addressing all inputs and outputs (energy and materials) involved in the production, use, and disposal of a product. This chapter describes the basic principles and methodologies used for determining and evaluating these life cycle impacts, with special emphasis given to plastics-related applications and materials.

The chapter limits itself mainly to discussing general production and disposal issues in typical plastics life cycles. Future chapters will give more examples and details about the chemical composition of specific plastics (Chapter 3), about how the environmental impacts of plastic products are related to the way they are used in specific applications (Chapter 4), and about how the design of plastic products influences environmental impacts (Chapter 5).

This chapter will make it obvious that because of their limitations, individual life cycle studies cannot normally provide final, definitive conclusions that apply to multiple products, market sectors, or companies' goals and needs. But it does argue that LCA studies are at least beneficial tools

Michael Tolinski and Conor P. Carlin. Plastics and Sustainability 2nd Edition: Grey is the New Green: Exploring the Nuances and Complexities of Modern Plastics, (29–66) © 2021 Scrivener Publishing LLC

for aiding the selection of materials for sustainability (Chapter 6). Briefly, the chapter is composed of these sections:

- Green principles for plastics (2.1)
- Life cycle assessment methods, metrics, limitations, and examples (2.2)
- The cradle-to-grave phases of plastics (2.3), with overviews on
    - raw feedstock materials for plastics: inputs and outputs
    - general service life cycle issues for plastics
    - plastics disposal, biodegradability, and recycling
- A proposed basic hierarchy for optimum plastics sustainability (2.4).

2.1   Green Principles

Any attempt to wade through the materials related to material life cycles and environmental sustainability is grueling, and developing useful conclusions is futile unless the process is anchored to certain guiding principles for sustainability. The basis of these principles is the acknowledgement that human activity makes certain demands on the earth's resources, creating wastes and disorder in the earth's system as a result. Inputted and outputted energy and materials flow through each stage of the production, use, and disposal phases of a man-made (or nature-made) object. Only nature seems to handle this complexity with any real skill. If people interested in sustainability have a shared goal, it would be to understand how to mimic the efficiency of nature in its use and reuse of the earth's resources. In fact, there is an entire philosophy dedicated to studying how man-made systems can mimic nature's efficient systems: biomimicry. Janine Benyus is a biologist who popularized the term in her 1997 book, *Biomimicry: Innovation Inspired by Nature* [1]. There are scores of examples of start-up companies that have developed technologies inspired by nature, including bullet trains based on kingfisher beaks, wind turbines modeled on humpback whales, and ventilation systems copied from termite mounds [2]. The plastics industry has also sought to mimic natural phenomena, as exemplified by the "Cora Ball". About the size of a softball, it is designed to capture microfibers from laundry, thereby preventing accumulation of microplastics in the water supply. The ball is modeled on the way coral filters the ocean. Made from recycled thermoplastic elastomer (TPE), it even has spiky tentacles, a testament to the ingenuity of the molder and the flexibility of the polymer [3].

To understand this complexity better, and how to achieve desirable goals, lists of green principles have been proposed by industry experts. Because plastics are composed essentially of chemicals in solid form, the "principles of green chemistry" have been one useful starting point. The twelve principles listed below are based on those from Anastas and Warner's important 1998 book, *Green Chemistry: Theory and Practice* [4]. Here, in quotation marks, are the authors' principles concisely paraphrased by the US EPA, and I have further attempted to put the principles into the context of plastics and their green issues:

1. "Prevention": Preventing waste is better than treating it or cleaning it up afterwards.

   Chemical reaction output wastes and unused inputs in polymer production must be controlled and minimized to keep a production facility economically and environmentally sustainable. And in creating plastic compounds, hard-to-handle power and liquid additives likewise are a potential source of waste from spillage.

2. "Atom Economy": Synthesis methods and processes should maximize the amount of raw materials used in the final product.

   The feedstocks and monomers used in high volume polymer production are of little value unless they are synthesized into polymers as efficiently as possible.

3. "Less Hazardous Chemical Syntheses": Synthesis methods should use or create non-toxic or low-toxicity substances.

   Some polymer processing, like other chemical processing, is not immune to the use of hazardous industrial materials. (One attraction of bio-based resins is the potential for avoiding most of these substances by using biologically based synthesis processes.)

4. "Designing Safer Chemicals": The toxicity of chemical products should be minimized.

   With plastics, the final "chemical product" is the molded or formed article, which is itself created from intermediary compounded pellets which are based on raw polymers and additive products. The toxicity of each end product must be minimized, especially because of the kinds of personal human contact applications that plastics are used for. In many developed countries, current and future regulations may seek to place the burden of proving a chemical's safety on the manufacturer before it can be approved for use, as regulators become more sensitive to plastics toxicity issues. REACH and RoHS compliance are key areas for company managers involved in environmental, health and safety (EH&S) roles. At a macro-level, deeper discussions about extended producer responsibility (EPR) have resulted in the shifting of responsibility upstream from municipalities to producers. This provides an incentive for producers to consider the environment in the design of their products [5].

5. "Safer Solvents and Auxiliaries": The use of solvents and other agents in production should be minimized, and solvents should be as safe and non-toxic as possible.

Solvents are crucial in most chemical processing and manufacturing processes (even if the solvent used is simply water). Their control and minimization, and the need for less hazardous solvent alternatives, are goals for plant operations as complex as polymerization processes and as basic as mold cleaning.

6. "Design for Energy Efficiency": The energy required for chemical processes should be measured, evaluated, and minimized.

Plastic processing typically requires heating materials and feedstock, plus mechanical energy for mixing, melting, and molding plastics. Various means can be used for minimizing this energy use (see Chapter 7 for some coverage of this issue).

7. "Use of Renewable Feedstocks": Raw feedstock materials should be based on renewable resources, when possible.

Obviously this is a major topic of this chapter and book. With plastics, the phrase above – "when possible" – is extremely difficult to decipher, since it can mean different things to different companies, products, and market sectors. Each company must decide when plastics based on natural materials are appropriate and practical to use.

8. "Reduce Derivatives": Multiple steps in chemical processing typically create additional intermediaries or wastes – but the amount of derivatives and steps should be minimized.

More complex polymers usually require more complex processing steps to synthesize them, increasing the likelihood of excess derivatives and waste. Sometimes, a less complex polymer compounded with additives may do the job as effectively, though complicated compounded plastics likewise require multiple processing steps or inputs, plus their own derivative materials.

9. "Catalysis": Selective catalytic reagents are better than stoichiometric reagents.

   Polymer synthesis has for decades made better use of efficient catalysts, which have become critical for creating so many of the low cost plastics in use today.

10. "Design for Degradation": Chemical products should break down into harmless components at the end of their use-life, and not persist in the environment.

   The biodegradability of biological and fossil fuel plastics is a key (and often controversial) topic – with questions concerning how to measure and define the true biodegradability of plastics. Questions also concern what importance plastics biodegradability has for sustainability, compared with the value or impacts of recycling, incineration, or landfill disposal.

11. "Real-time analysis for Pollution Prevention:" Methods should be optimized to allow real-time, in-process monitoring and control to prevent hazardous substances from being released.

   Some plastics operations have the potential to release hazardous materials if their processes are not carefully controlled, as in the case of PVC, for example, whose production has been linked to the release of vinyl chloride and chlorinated dioxins. Even in PVC extrusion, an overheating extrusion die can burn the PVC melt, creating airborne hydrochloric acid – a painful accident, to say the least.

12. "Inherently Safer Chemistry for Accident Prevention": The potential for chemical releases, explosions, fires, and other accidents should be minimized by the choice of substances used in the chemical process.

   This is a difficult principle to adopt for plastics, since most polymers and their feedstock materials are inherently flammable, especially in their bulk forms as pellets, powders, liquids, or flakes. Many plastics additives likewise are particularly flammable in their bulk state before compounding. Fires occur at plastics processing plants fairly regularly, but they can be minimized via controlled processing and material handling practices and other precautions.

Most of the above guidelines will be addressed directly or indirectly later in this chapter and book. Some have more to do with the origins and synthesis of plastics, and others concern the end of plastic products' lifetimes – areas of particular interest when evaluating the environmental footprints of plastic products.

2.2 Life Cycle Assessment (LCA)

Life cycle assessment (or analysis) has mainly been performed on industrial materials and products since the 1990s. Simply put, LCA is a methodology or technique for identifying, measuring, and evaluating all the energy and material flows that result from making, using, and disposing of a target product or material. The resulting assessment allows manufacturers to identify the most important ways of minimizing waste, energy, and the overall environmental footprint of a product or group of similar products.

The use and development of LCA has been growing slowly, given its complexities and limitations. LCAs require a great deal of work to complete, even for simple products. Thus their objectives and assumptions must be compelling enough to motivate the use of LCA by researchers, manufacturers, and other organizations. General motivating assumptions may include the following: a) that the earth's ecosystem and climate are being damaged by waste and pollution from human activities (specifically from the production of the material or product of interest), and that this damage can be lessened; b) that the earth's basic resources, including fossil fuels and fresh water, are being overused and depleted by producing the material/product; and c) that a manufacturer's internal economic efficiency, environmental impact costs, or public image are less than optimal because of the ways its processes and products use resources or produce waste. Obviously the first two assumptions would motivate environmentally focused researchers; the last assumption could motivate nearly all manufacturing-related researchers, whether they accept arguments [a] and [b] or not.

Various life-cycle evaluation methodologies measure various costs or environmental impacts from creating and using products. LCAs often focus on the metrics of CO_2 production (global warming potential) and other consumption, waste, emissions, and toxic material flows. After acquiring data for these flows and evaluating their significance, decisions can start to be made about how to reduce problematic issues with the product or system's environmental footprint. Some life-cycle approaches focus solely on energy consumption; others analyze monetary costs, social costs, and stakeholder impacts.

In this chapter, the term "life-cycle assessment" is used in a general way to refer to the general concept behind methods labeled "LCA" in university and industry sponsored research (and, more frequently now, in marketing appeals). In most of these cases, the LCA method actually used can be critically evaluated if certain material/energy flows or impacts are neglected or undervalued. Thus, an LCA can often give rise to spirited debate.

The full explanation of LCA methodology can be quite theoretical and is beyond the scope of this book. Figure 2.1 attempts to capture LCA principles visually. In brief, first an LCA's goal and scope must be defined, as well as its metrics, allocation of inputs and outputs, and system boundaries. An LCA is normally based around a chosen "functional unit" that is meaningful and useful for making comparisons (that is, a measurement or quantity of some "thing" to be assessed). Essentially, the LCA method then breaks down the stages of the unit's (product's) life cycle, looking at the material and energy inputs and waste outputs at each life-cycle stage. Below are life-cycle stages normally of interest, which we have tried to place into context for plastics-related manufacturing:

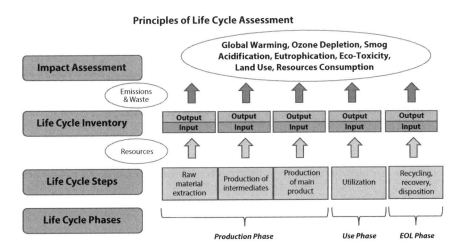

Figure 2.1 Illustrative diagram of life-cycle analysis (Source: Thinkstep).

- Raw material acquisition (such as oil or natural gas production, or in the case of bioplastics, biomass production).
- Material processing (the creation of monomers and reagents for polymerization, and subsequent polymerization and compounding to create bulk plastic materials).
- Manufacturing and assembly (the molding, extrusion, or other forming or conversion of bulk plastic compounds, plus any secondary operations).
- Use-life (the use of the plastic product by the consumer, including the energy for shipping or handling the product, and the effects of its use).
- End of life (the possible recycling or reuse of the product/material, before disposal).
- Disposal (the final resting place of the plastic product, whether it is landfilled, incinerated, or composted, including the disposal method's material/energy inputs and outputs).

When looking at this entire life cycle, a complete picture of a plastic's environmental impacts can be formed. An LCA might mainly emphasize the wastes involved in a plastic product's use, but sometimes they offer unexpected, dramatic results. LCAs have attempted to look at plastic materials as a whole – the "big picture." For instance, in a 2010 European study of plastics' total net effects on the environment, the overall use of plastics (in place of alternative materials) was credited with saving the equivalent energy of 50 million metric tons of crude oil per year (or 194 oil tankers), avoiding 120 million metric tons of greenhouse gas emissions [6]. However, less dramatic, small-scale LCA studies can indicate relatively subtle changes that can be me made to a single product to reduce its environmental impact – a potentially useful and worth-while exercise for a company. In 2018, the American Chemistry Council (ACC) worked with Franklin Associates to produce a comprehensive LCA comparing plastics used for packaging to substitute materials such as paper and glass [7].

2.2.1   Life Cycle Inventory (LCI)

The LCA process requires identifying all material flows at each stage of a product's production. This "life-cycle inventory" (LCI) lists and quantifies all materials that are present in the product, or that are required or produced in its processing, use, or disposal. This knowledge can be extremely valuable by itself for identifying and quantifying objectionable outputs from plastics production. An LCI can be valuable for minimizing plant emissions, avoiding harm to the environment and stiff monetary penalties for chemical releases.

When the LCI is as complete as possible, its flows are then converted into appropriate per-product basis data for use in the next LCA step – the life cycle impact assessment (LCIA). Here, various environmental impacts are calculated from the LCI outputs and inputs to assess the overall footprint of the product system or material unit. These impacts are then weighed and evaluated in the final interpretation phase of the LCA.

LCI data have been collected for many kinds of product systems, even an entire automobile. LCIs theoretically could compile very detailed data; for example, in an extreme case, LCI might identify the volume of water used in producing the volume of natural gas that is ultimately required for producing a unit of polymer. Such an example shows how deeply LCA can reach into examining a product's origins when strict system boundaries are not applied to its scope. It also shows the potential data collection challenges of LCI. Fortunately, public databases are available containing compiled data to aid people in performing LCIs (for example, information supporting plastics LCIs is available from the Association of Plastics Manufacturers in Europe [www.plasticseurope.org] and the National Renewable Energy Lab (NREL) [www.lcacommons.gov]).

2.2.2 LCA: Controversies and Limitations

Although they are intellectually stimulating and can be useful tools, LCAs require a great deal of interpretation for informed decision making. There are differences in LCA models used by different organizations, resulting in pitfalls associated with the use of these tools. For example, some LCAs can be over-interpreted or too specific, raising questions about their underlying data or assumptions; or they can be too comprehensive or standardized, limiting their particular usefulness. For the general public, the results of LCAs can be difficult to interpret or digest, even when authors attempt to quantify emissions in the form of trees planted or cars removed from the road.

Given their complexity and potential controversy, LCAs for plastics are often relegated to the realms of research or general product marketing, rather than used directly for practical product design and manufacturing. This may be because of the following limitations:

1. Traditional LCAs often do not always consider real-world infrastructure and monetary costs (materials and processing) as factors to help determine the real-life choices that can be

made in terms of raw feedstock inputs and disposal methods. Raw feedstock costs fluctuate, and their hard-to-anticipate price changes tend to determine, for example, whether oil-based or natural-gas-based feedstock is used in polyolefin production. An LCA could be performed for both cases for making comparisons about sustainability, but even then the more economical approach would be inevitable. Or consider the disposal phase of bioplastics. Composting is often argued to be the best "grave" for biodegradable biopolymers, as an LCA might also indicate, but this assumes falsely that the required large-scale industrial composting facilities actually exist in most municipalities. Here, the LCA's conclusion would have limited applicability.

2. Similarly, LCAs generally do not consider the availability, consistency, and stability of various raw material sources under consideration, and about how these factors determine material costs. In the case of bio-based polymers, key suppliers are still ramping up production, and most of the producers have limited capacity. Even if an LCA determines a bioplastic is the most sustainable choice for a product, its use may not even be feasible at the required volumes.

3. Unless carefully performed, comparisons of materials via LCA may not fully consider the quality or property differences between high and low environmental impact alternatives. Often, one element of a plastic's composition cannot be freely chosen from a variety of materials with varying environmental footprints. For example, a PVC composition with a high environmental impact additive, such as a standard plasticizer, may offer the desired properties, while another with a smaller environmental impact, such as a bio-based or phthalate-free plasticizer, may not.

4. LCAs do not factor in consumers' willingness or unwillingness to pay more for what is found to be a lower-impact product – or factor in whether there is enough consumer interest or regulatory pressure to make the LCA's findings relevant. Social factors may also make a low-impact material's use undesirable because of outcomes from its use, no matter what the LCA indicates.

5. Along the same line, it is difficult for users to judge what is absolutely sustainable in terms of environmental impacts.

There is relatively little agreement as to what level of emissions or waste is acceptably sustainable for a given type of product. A consumer who values his automobile or plastic water bottle a great deal might be willing to be somewhat liberal about what he considers sustainable waste and emissions levels. Moreover, the LCA method and interpretation may even be influenced by common public opinions about what the environmental impact of a product is. Thus an LCA's main strength is for making – as objectively as possible – relative sustainability comparisons between very similar materials, products, or processes.

6. LCAs, of course, do not factor in the marketing/public relations value of certain life-cycle inputs or outputs. Purists would not consider these as factors for an LCA, but they are unavoidable factors for decision makers in industry. For example, if a certain bioplastic shows fair performance in a credible LCA, one could argue that its overall score could be more enhanced simply by the public's feelings of goodwill about bioplastics. Conversely, in a study that determines PVC has the lowest environmental footprint for a certain product in a human contact application, the PVC's dominant ranking could be offset by negative public opinion about the use of PVC in these applications.

7. And finally, LCAs often cannot easily keep up with and take into account improvements in plastics manufacturing. For example, more efficient methods for producing polyolefin feedstock from sugarcane are now being produced at scale, changing bio-polyolefins' LCA results over just the past few years. Meanwhile, techniques for producing harder-to-extract crude oil and natural gas may become more polluting, increasing the environmental impact of traditional polymers in LCAs [8]. Yet traditional petrochemical processing is becoming more environmentally efficient, especially as processors are being pressured by regulators, the public, and environmental groups. Plastics producers are also becoming more adept at creating tougher, stronger polymers and compounds, requiring smaller amounts of plastic to make basic products, which trims their environmental impact. And recycling technologies are becoming more efficient as well, influencing recycling's end-of-life influence in an LCA.

Thus an LCA might be thought of as a hammer that is needed for reno-vating a house. Of course, many other kinds of tools are also needed to do this job, as well as some creative thinking on how to use the tools. By using various tools and strategic thinking along with the LCA hammer, the refurbishment of a product to reduce its environmental impact, or to make a different design choice altogether, is possible. Below, two examples of plastics-related studies better illustrate LCA's usefulness and limitations, as well as serving as an introduction to the issues discussed in future chapters.

2.2.3   LCA/LCI: Plastics-Related Examples

The following two summaries show how life-cycle studies were used to evaluate the environmental footprints of certain plastics. Besides providing thought-provoking results, each illustrates the choices that must be made about the scope, assumptions, and metrics of an LCA or LCI and how those choices can influence an evaluation.

2.2.3.1   PET and HDPE

Given the interest in expanding the recycling of commodity plastics pack-aging, it is useful to have LCI data close at hand. To this end, Franklin Associates (first in 2010 [9] and again in 2018 [7]) attempted to identify and quantify the key material and energy flows associated with post-consumer recycled high-density polyethylene and PET, to allow compar-isons of environmental impacts between the recycled materials and their virgin, pre-consumer forms. The authors created a usable study of lim-ited length by limiting their scope, identifying only the most important material/energy burdens for which they could acquire data.

The first report is full of stated assumptions (warranted by statistics about recycling) that are necessary in such a task. For example, the researchers had to estimate how much fuel is used by consumers to drop off recycled materials, how much fuel is used by curbside recycling collection vehicles, what percentage of PET and HDPE is contained in the collected materials, and so forth. The authors also calculated values based on the assumption that the recycled material could be expected to have two total "lives" (that is, a second life after recycling, followed by disposal). This determined how they should incorporate the environmental impacts of the virgin resin pro-duction when the recycled resin was originally produced.

Their detailed quantitative results were of the type that could be used in a full LCA. Here, some of their interesting results are summarized:

- The total energy for collecting, sorting, and reprocessing 1,000 pounds of post-consumer PET and HDPE was less than one-sixth as much as that required for producing virgin resin (which inherently possesses great energy content in its chemical composition).
- The greenhouse gas emissions from recycling, including emissions from material collection, were about one-quarter to one-third of those in virgin resin production.
- Solid waste production when recycling the post-consumer materials was higher than with virgin material production, because of the residuals and unusable materials produced by the sorting and reprocessing steps.

2.2.3.2   Bio/Fossil-Fuel Polymer Comparison

Researchers have also tried to compare different polymer types using life-cycle evaluations. Here, evaluating polymers using both LCA and the "Twelve Principles of Green Chemistry" might be expected to give similar, consistent results for each kind of polymer. But this is not necessarily the case.

A study performed by researchers at the University of Pittsburgh compared "green design" metrics and LCA rankings of multiple fossil-fuel-based polymers and biopolymers [8]. Common commodity and engineering polymers were ranked both on how well they satisfy principles of green chemistry and engineering and on their results from an LCA. Although the two approaches seem as if they would give similar results, they did not: based on green design principles, the researchers found that the biopolymers ranked the highest (lowest environmental impact); but in the LCA, simple polymers (polyolefins) ranked highest, followed by the biopolymers, followed by more complex polymers (PET, PVC, and PC).

Looking at this study in detail provides some answers on how the two assessments gave different results. Researchers considered two biopolymers, PLA and PHAs, each made by different processes (fermentation and bacterial synthesis, respectively). The authors also considered a partially bio-based PET along with a standard PET, plus six other standard fossil fuel-based polymers. The LCAs were reportedly performed in accordance with the ISO 14040–14043 series of standards, and only their "cradle-to-gate" production phase was studied, leaving out their use and disposal/recycling phases. LCA metrics included measurements of impacts from toxicity, pollution, global warming potential, and fossil fuel use in polymer production. Polymers were also compared via metrics based on green chemistry and green design principles, mainly related to their sourcing and production; these metrics included atom economy, density, toxicity, energy demand, feedstock transport, biodegradability, renewable/recycled content, and price [8]. In the results, the researchers found that the biopolymers performed well in several "green design principle" metrics. But they suffered in the LCA rankings due to their reliance on fertilizers and pesticides for plant-based feedstock production, causing higher impact scores in terms of ozone depletion, acidification, toxicity, and eutrophication. Simple traditional polymers like PP and PE performed better on the LCA because of their relatively straight-forward chemical processing. However, complex traditional polymers (PET and PC) were ranked lower than the biopolymers and polyolefins in both evaluations because of the multiple processing steps required for their production, and higher emissions. Interestingly, the partially bio-based PET ranked low using both methods, because of its multiple types of processing steps (and their emissions), which include both petrochemical and agricultural processes. "Switching from petroleum feedstocks to biofeedstocks does not necessarily reduce environmental impacts," the researchers concluded [8].

Such conflicting results reveal the difficulties of making environmental impact comparisons. These authors conceded that even this inclusive study had relatively limited scope, when considering that the impact effects of recycling, incineration, biodegradation and other factors were not included.

2.3   Plastic Lifetimes

2.3.1   The "Cradle": Polymer Feedstocks and Production

The above examples show how much the feedstocks of polymers influence their environmental footprints, given that the polymer chain itself stores so much energy. Raw materials for producing polymer feedstocks all have potential environmental impacts, whether through the risk of environmental damage caused by drilling for oil (as in the 2010 BP Deepwater Horizon disaster), or the inefficiencies of fertilizing and harvesting crops for biofeed-stock production, which currently still relies heavily on the use of fossil fuels.

2.3.1.1   Fossil-Fuel Feedstock Sources

Economic pressures will also affect the choice of fossil- or bio-based feedstock. Most polymer feedstocks are still based on the commodities of oil and natural gas. These commodities' prices ebb and flow, sometimes staying in a narrow price range for a long period (as oil did in 2010), or rising radically (in early 2011), or dropping dramatically (in 2014 and 2020). See Figure 2.2 below.

And of course, these are not renewable commodities; they are theoretically finite in supply and eventually will become much more costly and polluting to produce. However, at current production rates, oil is still expected to last several decades. In 2018, the US became the largest producer of crude, driving relentlessly by efficiencies related to shale oil and associated extraction techniques. Moreover, plastics use less than 5% of all fossil fuels, with some estimates showing that less than 1% of each barrel of oil is consumed by plastics production – less than the percentage for asphalt, lubricants, or coke [10]. And the other key fossil feedstock source for polymers, natural gas, remains in abundant supply, especially in the United States. Still, with plastics being used in greater quantities to help reduce the weight of automobiles, the subsequent decline in fuel demand is a topic of discussion in the board rooms of oil and chemical companies

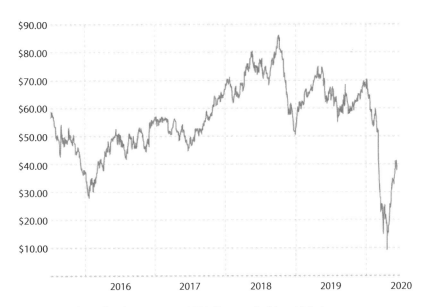

Figure 2.2 WTI crude oil prices 2015-2020 (Source: Seeking Alpha).

[11]. Normally, the oil-to-natural gas price ratio determines which raw polymer feedstock is economically favored (either using naphtha from oil or ethane from gas, for example, to create polyolefins).

Meanwhile, coal – the fossil fuel most commonly associated with pollution – is being turned into a feedstock in China for producing PVC, polyolefins, and ultimately other plastics precursors [12]. Coal-to-polymer technology may also be a hedge in coal-rich countries against high gas/oil prices. These "coal-to-chemicals" operations have questionable environmental impacts, but they tap a cheap resource in countries that have large supplies. The public's dislike of the damaging effects of coal mining, its water use, and pollutants are major disadvantages.

2.3.1.2   Bio-Based Feedstock Sources

Conversely, most current bioplastics are based on renewable agricultural "biomass" commodities such as corn and sugarcane, synthesized mainly by the free light of the sun and the greenhouse gas CO_2. But their price is affected by the weather, government subsidies, fertilizer costs, water consumption, land use competition, and global food supply trends.

Interestingly, until after the industrial revolution, biomass was the source of most industrial chemicals and materials. The earliest plastics were derived from plant materials and we have seen a renaissance of sorts in plant-based surfactants and household cleaners. Today, with petrochemical production dominant, modern, large-scale bio-feedstock production is still in a relatively immature state, making it difficult to compete with established fossil fuel processing in terms of cost and energy consumption.

One development that could decrease the cost and inefficiencies of bio-based feedstocks is the "integrated biorefinery" concept. Here a single operation converts biomass into a variety of useful products, making use of its entire feedstream (as a petroleum refinery does) to produce ethanol, methane, oil, chemicals, and other plant products [13]. Such biorefineries could accept various kinds of biomass other than corn or sugarcane, such as wood waste, switchgrass, agricultural wastes, and other lignocellulosic plant materials, which then become the useful building blocks of polymers. Fermentation of plant-derived sugars for producing ethanol and the bio-polymer polylactic acid (PLA) for example, would be just one of the many processes in the biorefinery. But technical hurdles remain to making these processes and biomass handling and separation more efficient.

Biochemical processes are at least starting to be used on a larger scale to produce some common polymer raw materials. Syngas (hydrogen and carbon monoxide) from biomass can be used for producing methanol, olefins, and other chemicals. And various plant sugars are being converted not only into lactic acid for PLA but also into "green" polymer feedstocks such as ethanol, glycerol, and succinic acid, assisting in the commercialization of lesser known biodegradable biopolymers such as polybutylene succinate. The raw chemicals also can be used for creating ethylene, 1,3-propanediol, butanediol, propylene glycol, and other raw components for building traditional polymers.

A biofeedstock source of recent interest is algae, which absorbs water, sunlight, and CO_2 from the air and converts it into oily bio-mass with extreme efficiency. Algae has been called "the ultimate winner" in the bio-feedstock race, because the organisms grow rapidly and their production does not disrupt food production and markets. They are said to produce useful bio-oils at a rate per unit cultivation area that is fifteen times higher than with other biomass sources [14]. Concerns about water use and practical technical concerns remain, however, such as how to cultivate and harvest the algae efficiently, either in an open body of water or closed reactor system. Commercially, these ventures have a poor track record in terms of scale, with many promising start-ups flaming out.

Other biomass materials are being researched and developed as sources or direct components of plastics, with emphasis on using non-edible or surplus agricultural materials. Years of work may still be required before any conclusions about the long-term viability of some of the following concepts, but most have at least passed beyond the phase of "pure research":

- Surplus soybean oil, reinforced with lignum from wood, is being developed by the United Soybean Board into resin monomers for sheet and bulk molding compounds and resin transfer molding. One example commercial application is a panel for John Deere tractors [15].
- Arkema recently announced that they were going to build a world-scale production facility to manufacture the amino 11 monomer for Rilsan® polyamide 11 resins derived 100% from castor oil [16].
- Non-edible plant cellulose and cashew nutshell oil (cardanol) have been chemically bonded with other additives to create a 70% plant-based thermoplastic material, according to the developer NEC Corp [17]. The material is said to have multiple times the strength and heat resistance of PLA, and equivalent water resistance.
- Keratin resin made from poultry feathers, combined with a polyolefin, is being molded into biodegradable flowerpots for nurseries, making use of at least some of the

3 billion pounds of chicken feathers created by the US poultry processors per year (roughly 80% of which is sent to landfills) [18].

Whatever the biofeedstock source, the true bio-content of the resulting polymeric material must be confirmed if the product can be called bio-based. The ASTM standard D6866 can be used to calculate the actual percentage of bio-based organic carbon in a material by measuring the proportion of the isotope carbon-14 in the material, and comparing it with a standard. (In other words, the organic fossil carbon in a traditional polymer would be expected to be all carbon-12; the carbon from a 100% plant-based polymer, however, would be expected to have a certain proportion of carbon-14 that was created by solar radiation in the atmosphere.) Narayan [19] suggests that although there is no "magic number" as to what this bio-content percentage should be for a plastic to be called a bioplastic, a bio-carbon content of at least 25% would signal that the material allows significant CO_2-reductions, especially if the material is used in high enough volumes to displace traditional fossil fuel-based plastics.

One more factor complicates the choice of using a traditional or bio-based plastic (a factor to be discussed more in future chapters). The properties of a bioplastic may be lower than those of a traditional material, requiring a thicker cross section in the product to achieve the same product quality. Thus, sometimes fossil-fuel-based plastic products can be argued as presenting a smaller environmental footprint than the equivalent bioplastic product. Because less overall material per product is used, its net life cycle impact may be lower.

2.3.2   "Gate-to-Gate": General Plastics Use-Life Impacts

Various properties of plastics influence the material and energy flows associated with their use-lives, from the gate of the factory they leave to the gate of the disposal/recovery facility they enter. Some of these properties were mentioned in Chapter 1, and forthcoming chapters will focus on these issues in more detail; thus this section will be short, touching on key issues briefly.

The high strength-to-weight ratios and low densities of plastics, compared with other materials, make them mass efficient solutions in many uses. In various applications, their durability and special properties can lower the *total net* energies and emissions associated with their production *and* use, if their use prevents waste. Their properties, as well as the efficient production methods for commodity plastics, lie behind the results of a

European study that credited plastics with overall greenhouse gas avoidance and reduced energy use when compared with alternative materials in most common applications [20]. Additional studies over the past decade support these findings [7]. Some conclusions from the study are integrated below, helping to support the following key points:

- Transportation fuel savings: As they replace metal parts in vehicles, plastic automotive parts will be credited with at least a small proportion of the reduced fuel use and greenhouse gas emissions in future vehicles. The benefits of choosing plastics rather than other standard materials results in energy savings that far offset the energy costs of producing and disposing of the plastics themselves.

- Packaging: Plastic containers are lighter than glass containers, reducing fuel use when transporting products. Packaging plastics also generally require less energy to produce than other packaging materials. Plastic film and containers are an effective means of prolonging food shelf life, reducing the amount of food spoilage and waste that would otherwise result. Even just a small amount of food loss prevention (10–20%) saves multiple times more greenhouse gas emissions than are produced by the production of the plastics themselves. Thus, plastics' unique characteristics have essentially allowed the creation of a system of food handling and distribution that is difficult for other materials to compete with in terms of total energy use (though most of plastic packaging demanded by this system does end up becoming noticeable and durable refuse).

- Building and construction: Plastics' production energies are lower than those of other building materials, especially when looking closely at the areas where these materials compete, and calculating their total energy and greenhouse gas impacts over the long lifetimes of buildings. For example, wood-plastic composites (WPCs) are long-lasting materials for decking, fencing, siding, and numerous other applications. These durable plastic applications generally do not need to be replaced as often as common wood-based products, preventing the production of replacement materials and their corresponding energy use. Some WPC applications also make heavy use of post-consumer recycled plastics, reducing their energy footprint further.

Despite their total energy saving potential, plastics can create various unwanted material flows during their use. As mentioned in Chapter 1, phthalate plasticizers leaching from flexible PVC and BPA residuals in polycarbonate or epoxy can liners are unwanted flows to the human environment. Or, for example, poorly formulated plastics may add unwanted tastes or odors to packaged food. And sometimes a plastic can be poorly chosen or designed for a certain rigorous application in which the product fails prematurely and has to be replaced, adding to the application's total energy use and environmental impact.

2.3.3   The "Grave": Disposal, Recycling, and Biodegradability

The end-of-life options for used plastic products include landfilling, incineration, or biodegradation via composting, when possible. It is often possible for some kind of reuse via recycling, or reappropriation of a part into a new product. Disposal is by far the most common method for plastics overall, and this is truly a waste, since plastics are primarily frozen chemical energy that required numerous chemical changes to create. Thus the issue of recycling remains intellectually compelling, though technologically and economically challenging. Its effects, as with biodegradability, can be measured and evaluated over periods of time, aiding in life cycle analyses. Thus, in this chapter, emphasis will be given to these short term disposal/reuse options for plastics, because these options' environmental advantages and impacts have a direct, immediate effect on the life-cycle economics and choices that producers and users of plastics face.

2.3.3.1   "Permanent" Disposal?

Landfill disposal is mainly a local issue, with some regions possessing large amounts of potential landfill volume. It is often an inexpensive and easy way to send plastics back to the earth from whence they came. Where landfill space is limited, incineration can be an efficient way of capturing the chemical energy of plastics and other materials and converting it into useful energy, as long as the right technologies and emission controls are used. Still, incineration produces carbon dioxide, a greenhouse gas.

Landfill disposal of the fossil-fuel-based plastics and biodegradable bioresins compared in this book is a mainly long-term solution whose impact differs slightly for each group of plastics. Theoretically at least, biodegradable bioplastics should relatively quickly close the loop on the carbon cycle, making the carbon again available in the atmosphere for use in biological growth (note that this does not apply to fossil fuel-based biodegradable plastics). Because CO_2 is captured from the air as their raw materials grow, one could say that their atmospheric carbon is "temporarily sequestered" in a sealed landfill for some months or years. Any anaerobic biodegradation that occurs in a landfill produces methane, which is released back to the atmosphere as a very potent global warming gas, if it is not collected and/or burned off and turned back into CO_2 by the landfill operation. Unfortunately, landfills in the United States are not required to install gas collection systems within two years of burying waste, thus any methane produced from quick bioplastic degradation could escape into the atmosphere [21].

This is unlike the landfilling of traditional fossil-based plastics, whose slow degradation over centuries slowly emits small amounts of "new" carbon into the atmosphere – carbon that had been stored deeper underground for millions of years as fossil fuel. So in terms of closing the loop in landfill disposal, biodegradable bioplastics may have an advantage in terms of long-term, life-cycle impact. But they may present a distinct disadvantage in the short term, in that the methane from initial anaerobic degradation escapes before it can be captured, as North Carolina State University researchers have noted [22].

Similar issues are connected with incineration, which is a direct way of extracting useful energy directly from polymer molecules. Plastics burn hot in waste incinerators. If processed in a combustion system for efficient energy recovery, 98.5% of mixed plastics would be converted to thermal energy [23]. In terms of life-cycle accounting, the fossil carbon of traditional plastics is released as "new" carbon into the air immediately after heat is obtained from it (though the burning creates energy that otherwise would be created by other fossil fuels anyway). The carbon in burned bioplastics can, however, be said simply to be returning to the atmosphere from which it was taken only a relatively short time before during plant growth. Conceptually, burned biopolymers provide a kind of "solar energy" from carbon captured via photosynthesis, as in the burning of wood. But in reality, bioplastics' energy from incineration would mainly compensate for the extra fossil carbon-based energy used during biopolymer synthesis. Overall, in incineration, bioplastics would still seem to be favored over fossil fuel plastics, though additional life cycle studies on this issue would be helpful.

2.3.3.2   Biodegradable Plastics

The biodegradation of plastics has become a controversial issue, probably because until recently most commercial high volume plastics were not biodegradable, so no agreed upon standard existed for what the term really means. But the concept is clear enough: under certain conditions, micro-organisms consume the polymer, ultimately reducing it to simple, natural gaseous compounds, water, and biomass. The final carbon emissions product of aerobic biodegradation process is CO_2 (and in an anaerobic environment, it is methane). Any true biodegradation must happen in a reasonable amount of time – weeks or months, rather than the dozens or hundreds of years required for degrading traditional plastics. Ideally the process should happen in nearly any natural environment or in a landfill, but in reality, it often requires controlled composting conditions for the proper mixture of air, heat, and microbes.

One obstacle in the use of more biodegradable packaging is the lack of composting facilities. Composting is also a relatively picky process. Among the relatively limited number of industrial composting operations that do exist in the United States, many require incoming materials to meet proper specifications and standards (about half require specific ASTM biodegradability standards to be met). And most of the composters would like more visible labels on products that identify certifiably compostable materials, because they often receive material that fails to break down as expected [24]. Moreover, some industrial composting firms may not wish to accept the materials without accompanying food waste in the waste stream, since food waste is more valuable in a composting operation than packaging waste.

To be composted or to simply fulfill the basic promise of their name, biodegradable plastics must first meet an appropriate definition of biodegradability. Being called a bioplastic is no help, since many bio-based plastics are not biodegradable. Showing that a material is truly biodegradable first requires testing it successfully against an accepted standard. There are a confusing number of accepted ISO, EN, and ASTM standards for measuring biodegradability: specification standards providing pass/fail guidance on making claims for biodegradability (ISO 17088, EN 13432, and ASTM D6400, D6868), and test methods for quantifying the degree of bio-degradability possible: ASTM D5988 (in soil), D5338 (in compost), D5511 (in anaerobic digesters), and D5526 (in accelerated landfill conditions), to name just a few. These provide percent-degraded results.

Along with bioplastics that are inherently biodegradeable, traditional plastics are being formulated with "oxo-biodegradability" additives that are designed to promote the oxidation and degradation of the polymer after a finite service life. Almost since their inception, however, these materials have been under close scrutiny due to the lack of clarity around their final decomposition. Some additives promote photodegradation as the initial degradation step, which starts after indoor products (such as plastics packaging), are exposed to outdoor sunlight (when littered, for example). The oxidized polymer chains break down into smaller units which can then be further degraded when exposed to microorganisms. An ASTM standard, D6954, is a three-tiered testing guide for measuring the property degradation, biodegradation, and ecological impact of oxo-degradable plastic formulations. This standard contains individual criteria that oxo-biodegradable material suppliers should cite their materials as passing, though the full standard requires extensive characterization of the material's degradation behavior.

Standards are important, since experts argue that unless it can be shown that a piece of plastic is converted by microbes quickly and completely into carbon dioxide, water, and harmless bio-mass, the term biodegradable should not be applied. Physical property losses and other basic changes in the material are not enough for extrapolating data to meet a claim of biodegradability. Otherwise, only partial degradation might result from the use of degradability additives; in these cases, claims of biodegradability would be misleading at best, and harmful at worst. This is because small pieces of partially biodegraded plastic can attract and concentrate toxic chemicals already in the environment, such as PCBs and DDT. If these small pieces are then eaten by birds, fish, or other creatures, it brings the toxins into the food chain and can cause irreparable harm.

This concern especially relates to the oxidation/degradation caused by the "oxo" additives. Here, all the degraded material may not be consumed completely or quickly by microbes at first, though the plastic compound is designed to degrade eventually over months or years. The key word here is "eventually" – questions remain about whether the partially degraded particles of these plastics accumulate in the environment or in the digestive systems of animals, rather than fully degrading.

More research is needed on what residual materials do remain after real-life outdoor degradation of various compounds and forms of plastics (including thick-section products) – and whether the residuals pose a special threat to wildlife. Despite claims from experts and producers of oxo-degradable materials, major brands have endorsed a total ban on the products [25]. In 2018, perhaps with the precautionary principle in mind, the European Commission began a process to restrict the use of oxo-plastics in EU countries.

2.3.3.3   Recycling

Source reduction, or minimizing the amount of resin per product, is the key way of reducing the life cycle impacts of plastics. But ultimately at least some polymer must be used in a product, and if current practices prevail, most of it will continue to be disposed of at end of life, the full value of its chemical bonds lost forever. Thus, the recycling of most plastics will remain an important goal, despite the particular challenges recycling presents.

To flesh out the context of sustainability for both fossil-fuel- and bio-based plastics, recycling is discussed at length in this chapter's final subsection, and in later chapters. There is at least one complex and important reason for focusing on recycling in detail: recycling is a key life-cycle factor that complicates the "fossil/bio" plastics issue. If plastics recycling were more practical and efficient, it might tend to deflate some of the pressure on the industry to develop bio-based resins. Ideally, highly efficient recycling of traditional plastics would improve their LCA rankings, lessen their environmental impact overall, and reduce attention on the problem of waste plastics. Optimally efficient recycling would be better able to answer public concerns about global warming, fossil fuel depletion, and landfill/incineration concerns. Inversely, a lack of comprehensive recycling (close to what we have now) puts pressure on the plastics industry to develop alternatives, or possibly will lead to more extreme regulations limiting plastics use, such as the shopping bag bans now being enacted. Moreover, special recycling issues with bioplastics are immediately relevant to their development – including questions about how or whether they can be recycled, about how they affect the quality of the conventional recycling stream, and about the economics of recycling overall (these issues are discussed more below).

Why recycle? For our plastics-dependent society, there are many reasons to continue pursuing recycling:

- Plastics' low density and flimsy qualities belie their relatively dense energy content and strong molecular bonds and structures; the waste of this energy through permanent disposal is simply poor systems design.
- The United States and Europe have recycling facility capacity that can process more post-consumer materials, though much of this material until recently had been sold and sent overseas (this amount is determined by the price differentials of recycled and virgin materials, at home and abroad).
- Plastics producers have developed technologies for incorporating more recycled material into its original application (rather than just downcycled into lower value applications like fibers and strapping), and there is still room for improvement.

- Consumers have developed a basic positive awareness about plastics recycling – their behavior is a critical part of the plastics recycling process – and this awareness could easily be lost in a generation if emphasis on recycling were reduced.
- Consumers are also open to buying products that contain recycled materials – as long as the price is right.
- And apart from citing percent-recycled figures (which are currently low for plastics and unimpressive), recyclers can cite life-cycle metrics that more persuasively show the potential value of efficient recycling, such as energy savings and greenhouse gas avoidance. But this can sometimes be a tough argument to make for plastics in practical reality, and the bar for recycling can be relatively high. Researchers have estimated that in the case of PET bottles, at least 50% of the collected post-consumer material must be transformed into usable recycled material in order to reduce PET's carbon footprint below the footprint created by simply landfilling all the bottles [26].

However, plastics production continues to outgrow plastics recovery, in the United States at least. This has helped keep the recycling rate of all plastics in the United States extremely low, at about 10%. And the rate for plastic recycling's star performer – the PET container – remained well below 30% in the years following 1997, which was roughly the start of the period when PET water bottle use started increasing. China's National Sword policy laid bare the extent to which US recycling data was dependent on that export market. For plastics other than PET and HDPE containers, the recycling rate in the United States is well under 10% [27, 28].

Along with consumption trends, the percentage of plastics recycled is linked to the fluctuating prices and costs of recycled and virgin materials. This makes the recycling industry in the West unstable and seemingly always in flux. Plants and companies shut down or go bankrupt, as others are being created or expand. The plastics recycling sector has also limited itself mainly to commodity plastics, which provide the highest, most consistent volumes of post-consumer material. This may be changing, slowly, as new chemical-based methods of recycling are commercialized; these can reduce multiple plastics into their chemical building blocks. So while some companies can survive by reprocessing lower-value plastics like PET or PE, other companies such as MBA Polymers have developed ways of separating and reprocessing higher-value, post-consumer engineering plastics, such as polycarbonate and ABS, in all their various grades and forms (extruded, injection molded, flame retardant, and so forth).

To some, the economic pitfalls of plastics recycling and low recovery rates may seem to indicate that it is not wise to expect recycling to flourish in developed countries with efficient petro-chemical industries. But it is also possible that the infrastructure and markets for recycling have not developed past a tipping point after which they become strong and stable. This may require intervention outside of overall market forces, as the US government and other governments often consider and do. Laws requiring deposits on plastic-bottled beverages is an obvious example, and this does increase collection rates, though it has traditionally been opposed by retailers and producers who dislike the associated costs and effects on consumer purchasing decisions. In the decade since the first printing of this book, for example, industry and government groups have worked together to develop preferred purchasing programs that favor environmental choices in federal or state procurement programs [29]. Demand-side initiatives are seen as a way to spur signals that will lead to private investment in technologies and commercial ventures that can supply new markets. Such interventions arouse ideological approval (environmentalists) and opposition (free market advocates), though one could argue that our oil-dependent economy does not account for cost externalities associated with fossil-based extraction and conversion.

The mechanics of recycling: Plastics recycling is technologically complicated and heavily dependent on the quality of the incoming material. Post-industrial scrap plastics are typically "clean" materials, consistent in form and quality, and they are commonly reprocessed, traded, and reused across the industry. Post-consumer plastics are another matter.

Mechanical recycling has several steps. Post-consumer plastics are collected curbside or from bins and are sorted from paper, metals, and other elements in mixed municipal waste/recycling streams. Much processing remains just to recycle one grade of one specific kind and form of thermoplastic, such as a blow-molded HDPE food or beverage container. In mechanical recycling operations, materials made from off grade or unknown plastic compositions must be removed from the stream, by hand or automatically. The foreign polymers are considered contaminants and must be separated cost-effectively; for this, high-speed sorting automation that uses density flotation separation or infrared spectra methods is important. After separation, initial grinding of the "good" recyclate aids in the removal of labels, metals, and other contaminants. The plastic stream then undergoes more grinding and washing; a step such as melt filtration purifies the material more by screening out contaminants from the melted material. Only then can the material be extrusion compounded with additives to counteract any degradation of the product that has occurred from heating or oxidation, and to prepare it for its next use.

Mechanical recycling is complicated by inconsistencies in viscosity, melt flow, and molecular weight of the incoming materials, and by variations in filler or additive content. The degree of material sorting and purification achievable by the process determines whether the recycled plastic can be used in new food/beverage packaging, rather than its more likely destination as fibers, strapping, non-food containers, construction materials, outdoor furniture, or the like. Mechanical methods are highly commercialized with many variations and new technologies that increase recyclate yield and quality (some are discussed in Chapter 7). But they work best only with a consistent, fairly clean stream of incoming plastics of known types, free of unwanted fillers, additives, or pigments. The greatest life-cycle improvement would be for 100% of a new PET bottle to be made from old bottles, but only a limited number of companies have been able to achieve this. Overall, mechanical recycling systems are based primarily on "…'chop and wash' processing technologies that were not designed to deal with the complex and highly variable plastics coming through our waste streams today" [30].

Other recycling operations reduce waste plastic back down to its original molecular constituents. The processes may be thermal or chemical. Thermal methods (thermolysis or pyrolysis) heat waste plastic in a controlled atmosphere, breaking it down into its basic chemical constituents that have industry value. Thermal methods are used for reclaiming common commodity addition-type polymers made up of chains of one chemical monomer, such as polyolefins (PE and PP), polystyrene, and PVC. Chemical options typically use solvents or reagents to "depolymerize" polymers into monomers or short chain oligomers, which can be used to create new polymers. These methods reclaim condensation-type polymers made up of two or more different molecular monomers, such as PET, polycarbonate, and nylon.

Thermal and chemical methods are less commercialized than mechanical recycling, and they are chemical- and energy-intensive. But they can produce valuable feedstocks, or at least usable petro-chemicals or fuels, from mixtures of waste plastics that are difficult or impossible to separate and reclaim mechanically. The current landscape for advanced recycling technologies has been mapped in great detail by Closed Loop Partners [30]. The authors identify 60 companies that are working on purification, decomposition, or conversion technologies. In one key finding, they note that it takes an average of 17 years to reach growth scale. Given the massive investments in virgin resin production (enabled by fracked shale gas),

LCA Study Metadata

Scientist and author Chris DeArmitt summarized the findings from over 50 LCAs and scientific papers in his website, www.phantomplastics.com. He started the project in response to the proliferation of opinion-based social media discussions that has led to confusion among the general public about plastics and the environment.

"It all started with a Google search. I looked for the terms "plastic bag lifecycle analysis" and "plastic bag LCA" because lifecycle analysis is the only internationally accepted standard to determine what is good and bad for the environment. GreenPeace uses it and so do governments and major companies. It's expensive and includes everything from cradle to grave including all the "inputs" (raw materials and energy), and "outputs" (emissions to the air and water, by-products and wastes disposal) for making a product. Because it's so expensive, I wasn't sure I would find any, but I was happily surprised to find some right away. You can type "LCA plastic bag" into Google and find many of the same hits I did."

Though Chris's work is data-driven and the LCAs he reviews show a strong consensus toward the environmental benefits of plastics vs. alternative materials, humans are not entirely rational beings and are not always swayed by facts. Rather, emotion, images, and narrative play a strong role in how we view the world around us. We all have our own biases and, in a market-driven, consumer-led economy, these biases become preferences to which companies respond. One only needs to think about how a single, viral image of a sea turtle led to a massive groundswell of support to ban this ubiquitous item.

market signals do not reflect future benefits that could be realized by larger investments in some of these companies. Advanced recycling is covered in more detail in Chapter 7.

2.3.3.4 Limitations and Challenges

Along with the cleanliness and overall quality of the incoming material, plastics have several particular qualities that interfere with their economical recycling. Unlike a steel can or glass container, which typically can simply be melted in a furnace at high temperatures without much damage to the material, plastics are organic materials that are degraded by repeated exposure to extreme conditions. And most are also relatively low-value-per-volume materials, compared with denser metals. This, plus the relative inexpensiveness of virgin polymers, weakens the economic drive for recycling in terms of relative costs vs. gained value. A number of related limitations or hurdles for recycling are addressed below.

1. **Low yields.** Conventional recycling cannot reclaim all collected plastic materials. Estimates vary, but roughly only half of the plastics in the municipal recycling stream may make their way into actual plastics recycling facilities. These recycling streams are typically mixtures of chemically incompatible polymers that cannot be reprocessed together because of differing melting temperature, melt viscosity, or molecular structure. These streams usually contain polymer types for which no recycling infrastructure exists, as well as plastic products that contain other materials that cannot be separated, plastic film or sheet that cannot be economically reclaimed (especially when it is contaminated), highly filled or colored items, or unusual polymer blends. Thermoformed or injection-molded products are also commonly rejected by recycling facilities that cannot accommodate them because of their melt viscosities or additives. And polystyrene foam presents difficulties, especially because its low density makes the costs of handling it hard to justify.

   Of the plastics that are accepted by a facility, estimates range from 70 to 90% as the proportion that actually comes out of the reclamation process as usable recycled resin. Thus, there is plenty of room for improvement, and material selection and design changes can improve these overall yields (see Chapters 5 and 6).

2. **Material degradation.** The recycling stream may contain polymers whose molecular chains have degraded from use, heat, or ultraviolet light. The additives in these materials designed to protect them may have been consumed, leading to reductions in the average molecular weight of the used polymer from breaks or "scission" of the polymer chains. Thus the incoming material has reduced properties even before it is reprocessed. By remelting material in the filtering

or pelletizing steps of recycling, the energy of reprocessing damages molecular properties further. Compensating for this additional "heat history" requires the adding of new stabilizers and potentially even extra enhancing additives that connect and lengthen the polymer chains, restoring some of the lost properties.

3. **Same polymer, different form.** Because of the material property requirements for a specific plastics forming process, recycled material from products originally made with that process are favored. So for the injection/blow-molding process for making PET bottles, recycled PET bottle resin is the only feedstream that is usually desired, because its molecular weight distribution and chain-length provides the necessary flow properties for the complicated process. For this reason, until recently at least, thermoformed PET products or other sources of PET were segregated from the bottle PET stream, even though the two clear packaging materials are similar. Thermoformed PET packaging is expected to grow to about half the level of PET bottles used in North America, creating a huge additional stream of recyclable PET material. But difficulties remain in processing thermoformed and bottle PET together, including viscosity differences, sorting difficulties (for example, thermoformed PET clam-shells can be difficult to sort out from PS, PVC, and PLA clamshells), label/adhesive issues, and additives issues. That said, investors have stepped up in this arena, responding to innovative policy developments in areas like California [31].

4. **Contaminating constituents.** Components in the product's composition that hinder recycling could include many different things, for example:

   o Dozens of kinds of additives, mineral fillers, and reinforcing fibers compounded into plastics often make materials containing them impossible to reclaim economically. Various plastic parts in the recycling stream may also contain unwanted halogen- or heavy-metal-based additives, requiring removal or higher-cost recycling methods. And it is often unclear what filler or fiber an old plastic part is loaded with (except in molded automotive plastics that are labeled with identifying codes). Even harmless common fillers such as calcium carbonate

($CaCO_3$) complicate matters; they are used in various degrees in recyclable PE products, and thus the amount of $CaCO_3$ actually in the recyclate is a variable to be accounted for.

o Bales of single-polymer plastics can often be contaminated with pieces of incompatible polymers. And parts having different polymer layers can also present trouble; a common-looking PET bottle may be blow-molded from material with multiple ethylene-vinyl alcohol (EVOH) barrier layers, for instance.

o Painted or coated plastics likewise are prohibitive or problematic, with paint-removal operations adding process costs to the recycling operation.

o And the recycled polymer may also have been contaminated or altered by material it contained or contacted during its service life, such as milk, label adhesive, or motor oil. An extreme example would be agricultural films, where the dirt collected with the film makes recycling very difficult.

Partly because of these reasons, much commonly recycled plastic is downcycled into material for less sensitive applications: 40–60% of recycled bottle PET ends up as fibers for clothes or carpet, and well over 50% of HDPE food/beverage bottles are turned into non-food bottles, pipe, or plastic building material.

A more recent and contentious contaminant issue concerns the greater amounts of bio-based and biodegradable plastics in the packaging stream. Biodegradable bioplastics, as well as traditional plastics compounded with additives that promote oxo-biodegradability, are likely to become more prevalent in conventional recycling streams of PET and PE. These polymers have the potential for significantly disrupting recycling operations. For example, a biopolymer such as PLA used in packaging is difficult to distinguish visually from conventional PET packaging. But it must be separated, because PLA is incompatible with PET in recycling operations, and its presence interferes with the most efficient sector of plastics recycling that currently exists. Thus, some have argued for a moratorium on PLA's use in bottles until this recycling conflict can be addressed.

The argument here is essentially that PLA adds unwanted social costs – not just costs for recyclers, but indirect costs to the public from PLA's effects on PET recycling. Meanwhile, processes for handling large amounts of PLA recyclate are being developed, with better sorting methods and a chemical process for using hydrolysis to break down PLA into lactic acid.

Traditional PET and polyolefins containing oxo-biodegradable additives are outwardly identical to, and compatible with, standard PET and polyolefins in recycling. However, problems may lurk with plastics containing many kinds of oxidizing additives coming on the market, each of which behaves somewhat differently. The degradable compounds are designed to break the polymer down after a certain period of exposure to room temperature, essentially giving it a shelf life. If the recycling process initiates this oxo-degradation of some material in the recycling stream, the additives could degrade molecular weight and properties to the point that it compromises the quality of the entire stream. Or, products made with the recycled products, such as strapping, might start degrading prematurely when in use and fail because of the additives. Unless these recycling and biodegradability issues are clarified through testing and labeling, retailers and consumers will soon become confused in understanding the acceptable way to discard these plastics. Nor will they be sure whether a degradable material is truly safe for animals and the environment if it is littered. Continuing disputes between bioresin producers, recycling trade groups, and oxo-degradability additive proponents have not (yet) led to a resolution of the confusion. Outright bans on oxo-degradable materials have been implemented in some European countries, with others studying the issue with greater scrutiny [32].

Limited material availability. It is true that many consumers do not recycle their bottles and packaging plastics (or anything else). Their behavior limits the amount of material available and the economic/environmental advantages of higher-volume recycling.

However, a large proportion of plastics that could be recycled lie outside the range of municipal collection. The automotive industry uses huge quantities of carefully engineered plastics in vehicle bumper fascia, doors, panels, and so forth – little of which is recycled at end of life. And consumer electronics typically have high-value, flame-retardant plastic housings, buttons, and connectors – very little of which is recycled in efficient processes. As mentioned in point 4 above, most of these materials have contamination issues which add costs to their recovery. Laws motivating the acceptance of electronics for recycling have taken effect in various countries and some US states; these may motivate improved recycling technologies.

5. *Consumer confusion – "people are people".* Of course, post-consumer recycling requires the cooperation of consumers. Compared with metal cans, glass, or paper recycling, plastics recycling requires some attention as to what is being put into the bin. Some curbside recycling guidelines require consumers to examine the resin identification number at the bottom of each container, with only certain numbers being allowed in the bin (with both consumers and recyclers assuming, or at least hoping, that the resin ID number on the product is factually correct). Guidelines might also tell the consumer to remove and discard the polyolefin caps from bottles. And sometimes consumers might even be asked to make judgments about whether a container is injection molded (the container's opening is as wide as its body: "unacceptable") or blow molded (a narrower neck opening: "acceptable").

A country such as Japan with limited landfill space has developed a social ethic for recycling, with strong rule enforcement for following these recycling guidelines. But in countries such as the United States, perhaps it is not surprising that some consumers simply throw all their plastics into the recycling bin without sorting them – or choose not to recycle at all. After all, human behavior is difficult to control, especially when the only direct benefits/rewards or costs/penalties for certain actions is a slight sense of moral rectitude or shame.

2.4   A Hierarchy of Plastics for Sustainability

Given all the complexities of life-cycle issues with plastics, some form of simplification is needed for guiding the next phases of discussion. A first step is to simply visualize what the basic goals should be when making decisions about which plastics are desirable in terms of their environmental footprint.

One visualization that has become popular is to plot material choices on a triangle or pyramid graphic that emphasizes their sustainability. In Figure 2.3, I propose a similar alternative: an inverted pyramid which emphasizes, at its top, not only the favorable environmental footprint of desired materials but also the high volumes in which these materials would have to be used in order to have significant effects on global sustainability. (Figure 2.3 is a stripped-down, idealized visualization; Chapter 6 provides more detailed, practical charts which are more useful for materials selection.)

Figure 2.3 proposes what is ideally wanted for plastics sustainability. Its top level emphasizes the future use of high volumes of simple, renewably sourced plastics that are also either easily recyclable and/or biodegradable. The next two steps down are where the industry is today. It can produce recycled and bio-based plastics in commercial quantities, and

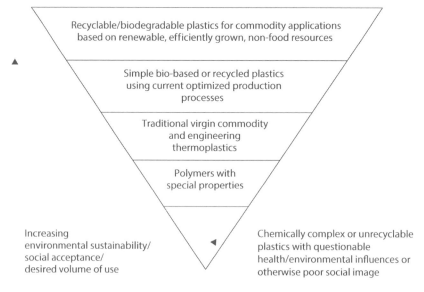

Figure 2.3 Inverted pyramid emphasizing goals in choosing plastics for sustainability.

it regularly produces mega-tons of basic commodity polymers with very efficient processes. At the next levels down are where traditional engineering polymers and specialized polymers belong – they should be used in minimal amounts where they are needed, allowing them to lend their own environmental benefits from their use, such as in improving the fuel efficiency of vehicles with lightweight automotive plastic parts. At the bottom, the least favored level of the inverted pyramid, are polymers that are unrecyclable or based heavily on chemical functional groups, including chlorine or aromatic chemistries, associated with pollutants or toxic materials.

With such an ideal hierarchy in mind, readers will be presented with details in the next chapter about specific plastic materials. Exploring this level of technical background will be helpful when making practical choices about plastics to minimize the life cycle impacts of a product.

References

1. Benyus, J.M., *Biomimicry: Innovation Insipired by Nature*, HarperCollins/Perennial, New York, 1997.
2. https://www.digitaltrends.com/cool-tech/biomimicry-examples
3. Grace, R., From Liquidating Mixed Plastic Waste to Capturing Your Washing Machine's Microfibers, in: *Plastics Engineering*, Wiley Publishing, June 2017.
4. Anastas, P.T. and Warner, J.C., *Green Chemistry: Theory and Practice*, Oxford University Press, Oxford, 1998.
5. Hannequart, J.P., EPR Schemes for the Benefit of LRAs?, in: *Assoc. of Cities & Regions for Recycling & Sustainable Resource Management*, Accessed via ec.europa.eu/environment, 2013.
6. Smith, C., European Union Parliament hears the 'green' side of plastics, in: *Plastics News*, http://www.plasticsnews.com, 2010, April 30.
7. Franklin Associates, Life Cycle Impacts of Plastic Packaging Compared to Substitutes in the US and Canada, Theoretical Substitution Analysis, prepared for the American Chemistry Council, April 2018.
8. Tabone, M.D., Cregg, J.J., Beckman, E.J., Landis, A.E., Sustainability metrics: Life cycle assessment and green design in polymers. *Environ. Sci. Technol.*, 44, 21, 8264–8269, 2010.
9. Franklin Associates, *Life Cycle Inventory of 100% Postconsumer HDPE and PET Recycled Resin from Postconsumer Containers and Packaging*, Franklin Associates, Prairie Village, KS, 2010, April 7.
10. Singh, B.B., Bio plastics for global sustainability [Conference presentation slides], ANTEC 2010, in: *Society of Plastics Engineers*, 2010, May.
11. Butler, N. and Tipaldo, E., Unlocking the Plastics Paradox. *Resour. Recycl.*, 2019, February 26.

12. https://www.reuters.com/article/us-china-coal-polyester-hengli-focus-idUSKCN24U0MA, July 2020.
13. US Dept of Energy and Office of Energy Efficiency & Renewable Energy (EERE), https://www.energy.gov/eere/bioenergy/integrated-biorefineries, Aug 2020.
14. Biron, M., Reducing the pressure on food crops with algae-based biopolymers. http://www.specialchem4polymers. com, 2010, September 28.
15. From petroleum to bio-based resins. (n.d.), Composites Manufacturing.
16. Arkema company press release, March 2020.
17. NEC Corporation, NEC develops high performance bioplastic with a high plant ratio by using non-edible plant resources (Press release). http://www.necam.com, 2010, August 25.
18. Verespej, M., Bioresin turning feathers into flowerpots, in: *Plastics News*, http://www.plasticsnews.com, 2010, July 28.
19. Narayan, R., Bioplastics: Next generation polymer materials for reducing carbon footprint, and improving environmental performance (Webinar), in: *Society of Plastics Engineers and Ramani Narayan*, Michigan State University, 2008, June 4.
20. Pilz, H., Brandt, B., Fehringer, R., *The impact of plastics on life cycle energy consumption and greenhouse gas emissions in Europe (Report)*, denkstatt GmbH, Vienna, 2010, June.
21. Wright, S., Biodegradable products in landfills may be harmful, in: *Waste & Recycling News*, http://www.plasticsnews.com, 2011, 2011, June 1.
22. Levis, J.W. and Barlaz, M.A., Is biodegradability a desirable attribute for discarded solid waste? Perspectives from a national landfill greenhouse gas inventory model. *Environ. Sci. Technol.*, http://pubs.acs.org/doi/abs/10.1021/es200721s, 2011.
23. Tsiamis, D. and Castaldi, M.J., Determining Accurate Heating Values of Non-Recycled Plastics (NRP), in: *Earth Engineering Center*, City University of New York, City College, March 2016.
24. US composters call for better bioplastics labeling, *European Plastics News*, http://www.plasticsnews.com, 2010, June 10.
25. https://www.foodbev.com/news/food-companies-back-initiative-ban-oxo-degradable-plastics/, 2017.
26. Cornell, D., Circular Economy and Virtue-Signalling: Consequences and Impossibilities. *Conference proceedings, PETnology Americas 2018*, Atlanta, GA, 2018.
27. https://www.waste360.com/plastics/us-plastic-recycling-rate-projected-drop-44-2018.
28. https://resource-recycling.com/plastics/2018/08/01/epa-u-s-plastics-recycling-rate-declines/
29. https://www.recyclingtoday.com/article/plastics-environmental-purchasing-guide/, 2018.
30. Report by Closed Loop Partners, Accelerating Circular Supply Chains for Plastics. 12, 2018.

31. Johnson, J., rPlanet Sees A Whole New World for PET Recycling, in: *Plastics News*, 2018, Feb 14.
32. Qureshi, W., Plastic and Environmental Associations Call for Ban on Oxo-Biodegradables), Packaging News.co.uk, 2020, June 2.

3

Polymer Properties and Environmental Footprints

This chapter begins the real process of comparing and contrasting individual polymers and plastic formulations in terms of properties and sustainability. (This process continues in Chapter 4 – "Applications" through Chapter 6 – "Material Selection".) Below are the essentials concerning the most common polymers in use – their basic properties, chemical footprint, processing requirements, and end-of-life options. The chapter will not cover the core basics about each polymer (its structural formula, for instance), which can be found in multiple polymer textbooks and online.

In particular, a focus will be put on details about the newest commercial biopolymers*, plus details about essential fillers and reinforcements, bio- and non-bio-based, that are used in plastic compounds. These details, plus

---

* Since the first edition in 2010, there have been many changes in the bioplastics industry. New companies have formed while others have pivoted to related areas, failed or closed. This section reflects the current landscape at the time of publication.

Michael Tolinski and Conor P. Carlin. Plastics and Sustainability 2nd Edition: Grey is the New Green: Exploring the Nuances and Complexities of Modern Plastics, (67–126) © 2021 Scrivener Publishing LLC

information in the next chapter about how individual plastics are used in real-life applications, are important to have when making material selection decisions about cost, performance, and sustainability. Below is an overview of the chapter's structure:

- Background on polymers and plastics (3.1), and "green principles" of interest.
- Traditional commercial polymers:
  - Common commodity thermoplastics: PE, PP, PVC, PS, and PET (3.2).
  - Common engineering thermoplastics: nylons, ABS, and polycarbonate (3.3).
  - Thermosetting plastics and conventional composites (3.4).
- Bioresins: Polymers of biochemical origin (PLA, PHAs, starch-based, protein-based, and biopolymer blends) (3.5).
- Conventional and bio-based additives, fillers, and fiber reinforcement (3.6):
  - Additives overview
  - Mineral fillers
  - Fiber reinforcement: glass, carbon, and plant-based fibers
  - Bio/nanocomposites

To support the process of making "greener" judgments about plastics, various polymers and polymer families touched on below are described in terms of their properties and environmental impacts from their synthesis, use, and end-of-life stages. First, some key concepts about polymers are reviewed that are critical to consider when comparing plastics in terms of environmental sustainability, and then individual polymers will be discussed. This discussion is by no means comprehensive; rather, it gives special attention to materials that are most important in the fossil-based/bio-based plastic debate.

3.1   Background on Polymers and Plastics

To plastics engineers and polymer scientists, trying to define "plastics" creates debate. Plastics are polymers, but not all polymers are plastics. Given the current global discussion about plastics in the environment and the potential ramifications for legislation, investments, and business sentiment, getting into the semantic weeds is not helpful. We proceed, therefore, with enough simplification to allow the non-scientist reader to continue, while acknowledging the complexity of the topic.

Polymers are the core materials that, combined with additives, make up what we commonly call "plastics." Composed of molecules that each contain thousands of repeating chemical units (monomers), polymers are synthesized from other synthesized or extracted chemicals, sometimes in several complicated steps. Polymer properties vary according to what chemical building blocks are used to create the polymer: generally, the more complex or expensive the monomers are, the more extreme the properties of the resulting polymer are. Although the exact synthesis process details are beyond the scope of this book, focusing on core information about each polymer still allows for basic comparisons in terms of sustainability and properties (made mainly in Chapter 6). Comparisons must be based not just on the number of processing steps and inflows/outflows in the production processes (and other "green" factors), but on polymer cost, strength, stiffness, density, barrier properties, impact resistance, and aesthetics.

To become useful plastic products, polymers require additives for protecting the polymer chain from heat, oxygen, or light damage, and for providing specific properties for an application. Some plastic compounds contain only a small total percentage of additives, while some, such as flexible PVC, may contain over 30% additives. Fillers are also added to displace the amount of expensive polymer that is needed to modify or enhance a wide variety of polymer properties such as impact resistance, stiffness, clarity, etc. These fillers may be added at weight percentages well over 30%. Reinforcements such as glass fiber or other kinds of fibers provide specific mechanical or electrical property enhancements when compounded with polymers for engineering applications.

Plastics are used in so many different applications for at least five main reasons: (1) polymer backbones made from various monomers can provide a wide range of properties; (2) the molecular structure of each kind of polymer can be controlled in production to produce specific properties; (3) at the polymer compounding stage, additives, fillers, and reinforcement can be very efficient and flexible ways to achieve desirable properties; (4) plastic product design and processing options allow great latitude for exploiting a material's strengths or hiding its weaknesses; and (5) the processing variables during the molding or forming of plastics can themselves be used to enhance a product's properties. The first three points are emphasized below.

3.1.1   Green Chemistry Principles

When considering plastic compounds in all their compositions and forms, selecting the right compound for a product is difficult to do, even when just comparing their basic material properties. Careful consideration of their environmental and human-health impacts complicates the equation. Fortunately for plastics, commercial polymers are usually composed of molecules that are too large or heavy in molecular weight (>10,000) to be easily transported across biological membranes, making them less prone to affect biological processes [1]. But of course, there are still many plastics additives and remnants or residuals from polymerization in the compound; these have a greater potential of being released into the environment and interacting with life processes.

In an effort to clarify each polymer's health and environmental impact, sections in this chapter will try to show how polymer synthesis and disposal processes relate to some of the "Twelve Principles of Green Chemistry", which were fully reviewed in Chapter 2. In particular, some of the principles most relevant to this chapter are selected and restated here:

- *Principle 3: "Synthesis methods should use or create non-toxic or low-toxicity substances."* Each of the polymer families described below requires various precursors or building-block chemicals for their creation. Some are built on simple, relatively low-impact chemistries; others require more complicated precursors, which may themselves be based on chemicals that are controlled or hazardous materials. Favored polymers would be those made from simpler or less-hazardous synthesis elements.
- *Principle 4: "The toxicity of chemical products should be minimized."* For plastics, the chemical product is essentially the final molded or formed article itself, made from polymers and additives. Plastics are used in the most personal of manufactured products; thus the toxic or environmental impact of the end product must be minimized. Admittedly, this is sometimes difficult to judge because of public controversies and "data wars" between parties who disagree about the health effects of certain plastic chemistries. This book will take the approach many manufacturers do by simply disfavoring the use of plastics which have major public

perception problems, or which multiple studies suggest carry an excessive risk of causing negative health effects from their production or use.

- **Principle 6. "The energy required for chemical processes should be measured, evaluated, and minimized."** Any attempt to quantify total energy required to produce different polymers is beset with challenges, but it doesn't mean engineers and chemists will stop trying. The current LCA debate is perhaps the next iteration of a discussion from decades ago about how to calculate energy requirements in a shifting environment. New process technologies, new resins, and fundamental changes in feedstock, i.e. petroleum vs. natural gas, combine to create a thorny problem. Researchers have tried to combine both the inherent fuel-energy feedstock content of fossil-fuel polymer, plus the total synthesis process energies required for producing each polymer. Biopolymers such as PLA and PHA can appear very energy-efficient to produce, but the accuracy of general numbers is difficult to verify, considering all of the possible material/process factors that could be included in the calculations. Moreover, the numbers do not include energies required for compounding and forming polymers into plastic products, and these energies vary according to each application, forming process, and polymer.

- **Principle 7: "Raw feedstock materials should be based on renewable resources, when possible."** Methods and systems for creating traditional polymers are, for now at least, overwhelmingly tied to oil and natural gas production. This is true economically and geographically, with more new plants for producing polymers now being located in the United States thanks to the discovery and development of shale deposits that has conferred a large strategic advantage on US-based producers and created a trade surplus in resins. However, most of the biopolymers mentioned below are based solely on biochemical fermentation or other biological processes that use biomass, mainly plant materials, as a feedstock source. And traditional polymers can use components made from renewable feedstock for their precursors, effectively increasing their "bio-basis" percentage to significant amounts.

Table 3.1 Representative property ranges of polymers covered in this chapter.

Polymer	Density	Tensile modulus (GPa)	Tensile strength (MPa)	Elongation at break (%)	Notched Izod impact strength (J/m)
Polyethylene (LDPE)	0.92	0.2–0.3	8–30	100–900	500 to no break
Polyethylene (HDPE)	0.96	0.8–1.5	28–32	10–300	150 to >1000
Polypropylene	0.9	1.1–1.5	25–33	50–300	70–150
Polystyrene (general purpose)	1.05	3.3–3.4	34–36	1–2	20 or less
PVC (unplasticized)	1.4	2.8–3.0	50–55	60	70–80
PET	1.4	3.0	50–75	50–300	20
Nylon 6/6 (PA 6.6)	1.1	2.9	60–80	60–80	50–110
ABS	1.04	2.1–2.4	20–55	8	200–350
Polycarbonate	1.2	2.3–2.4	60–110	100–110	700
PLA	1.24	3.0–3.5	26–144	3.5–8.1	16–144
PHA	1.3–1.4	0.1–3.0	19–25	4–450	26–37

Note: Ranges for traditional polymers based on values from multiple sources; value ranges for PLA and PHA are from NatureWorks datasheets [2] and Mirel (Yield10) Bioplastics provisional datasheets [3], respectively.

- *Principle 10. "Chemical products should break down into harmless components at the end of their use-life, and not persist in the environment."* Although not all bio-based plastics are biodegradable, the relative degradability of all polymer systems can be compared, and the ways they do (or do not) break down. The advent of the circular economy, however, has brought into question whether and how chemicals and plastics are re-used. Other end-of-life factors for products made of each polymer, especially recycling, will be discussed in the subsections below.

Of course, apart from these green concerns, a material's mechanical properties determine whether a polymer can be used for an application in the first place. Table 3.1 gives some representative mechanical properties for the polymers covered in this chapter, showing how property ranges of key biopolymers (PLA, PHA) overlap with those of traditional, fossil-fuel-based polymers applications such as food packaging, such as gas transmission and barrier properties. Often these special properties are *the* determining factors in material selection; for example, PET, unlike PLA, has the gas barrier properties that allow its use for carbonated beverage bottles. Heat resistance is also a critical property. For instance, PET typically cannot be hot-filled with food or beverages, while PP can.

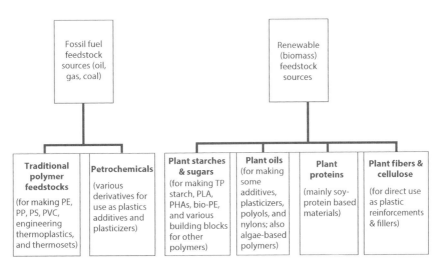

Figure 3.1 Categorization by feedstock origin of the fossil fuel- and bio-based materials covered in this chapter (with guidance from [4]).

The sections below emphasize factors that define the eco-footprint of each polymer, while not ignoring their key properties and applications. They cover the most commonly used traditional and bio-based polymers and additives, which are simply categorized according to feedstock origin in Figure 3.1. Particular attention is given to polymers used in high-volume commodity applications, because the total environmental footprints of commodity polymer applications are ultimately what influence overall plastics sustainability issues – more than even the combined effects of plastics in specialty, niche, or low-volume engineering applications.

3.2 Common Commodity Thermoplastics

The key commodity plastics below, combined, make up by far most of the total volume of plastics produced and used (see Figure 3.2 for breakdown of prime plastics into major thermoplastic categories). Thus, extra details are given here about their production and disposal impacts.

3.2.1 Polyethylene (PE)

PE is a polyolefin – a polymer built from repeating units of simple hydrocarbons. PE is the most versatile and popular polymer chemistry for low-value applications, and it comes in many different grades and forms.

3.2.1.1 Synthesis

In general, PE is created through addition polymerization, where catalysts drive the bonding of one ethylene unit to another to create long polymer chains. Over the last few decades improvements in single-site catalysts have allowed the synthesis of PE grades with very specific and consistent molecular structures chain lengths, and chain branching– and thus very predictable physical properties. PE requires the least energy to produce of all traditional polymers (see Table 3.1); its basic, conventional production synthesis steps are simply:

1. Ethane is separated from natural gas or obtained from crude oil refining.
2. Steam cracking converts ethane to ethylene.
3. Catalysts are used to polymerize ethylene.

PE is also created in production quantities from feedstock derived from sugarcane. This PE is chemically identical to fossil-fuel PE. The process

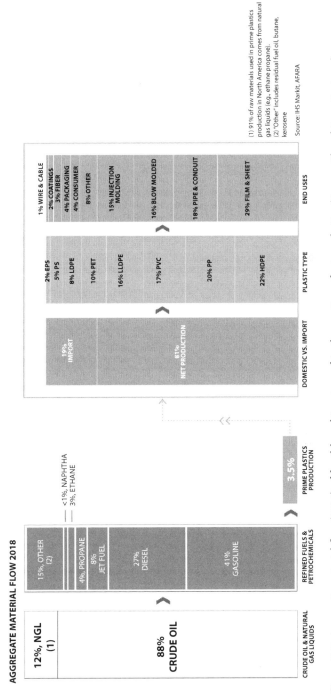

Figure 3.2 Aggregate material flow (2018) of fossil-based inputs used in the creation of prime plastics (*Source:* Closed Loop Partners).

converts cane sugar into ethanol (similar to the process for using corn starch to create ethanol), which is then dehydrated into ethylene [5]. The steps can be summarized as:

1. Raw sugarcane material is pressed to separate its sugar juice.
2. Microbes digest the sugar via fermentation, creating ethanol (and CO_2).
3. The ethanol is dehydrated, requiring catalysts and heat to create ethylene and water.
4. The ethylene is polymerized with catalysts to become PE.

When working with plant starch (such as corn) as a biofeedstock, the process is similar, except that precooking and enzymes are first needed to convert the starch into a sugar syrup for use in fermentation.

From a green principle/LCA perspective, the process of producing PE from biofeedstock can look attractive. After all, to produce each kilogram of bio-based PE, over two kilograms of CO_2 are removed from the atmosphere by the living plant during photosynthesis. However, waste materials and some CO_2 are returned to the biosphere during bio-PE's production. And when focusing on the conversion of the plant's sugar (glucose) molecules into ethanol, bio-PE looks less efficient: About half of the glucose is lost as CO_2 during the creation of ethanol. After the ethanol is dehydrated into ethylene for polymerization, the net theoretical yield of turning glucose into PE drops to 31% [6]. Nonetheless, the estimated ranges of greenhouse gas (CO_2 equivalent) emissions during bio-PE production are still lower than the emissions from standard fossil-PE production [7].

This bio-PE process has been commercialized and is sold under the name "I'm green™ Polyethylene". In 2010, Brazilian petrochemical company Braskem SA opened a 440-million-pound-capacity plant for producing bio-PE from sugarcane, a crop which Brazil also relies on as a major ethanol fuel source. Growing the crop for ethanol production reportedly requires less than one percent of all of Brazil's arable land, and it does not damage the Amazon rainforest ecosystem, because sugarcane cannot be grown in that climate [8]. In 2018, Braskem launched a new resin made from sugarecane, ethylene vinyl acetate copolymer (EVA), further strengthening their bio-based credentials and product portfolio [9].

Nonetheless, the fact that bio-PE's feedstock currently comes from a cultivated food crop is still an undesirable element. Thus there is motivation to create cellulose-based ethanol using plant stalks, wood waste, or nonfood plants that are easy to grow as inputs. In these processes, size reduced, pretreated cellulosic material must be hydrolyzed into fermentable sugars by enzymes or acid. Common pretreatment methods include "explosion methods" using saturated steam, or related methods, to expose the plant fibers' cellulose for conversion into sugars [8]. Given the more complicated streams of materials in the process, the process of fermentation is also more complicated, requiring careful combinations of bacterial strains.

3.2.1.2 Structure and Properties

Depending on its polymerization process, PE grades can have selectively designed structures and densities, with varying degrees of branching of polymer chains off each main polymer backbone. Hence the basic types of PE: high-density polyethylene (HDPE) with limited short chain-branching, allowing high crystallinity; low-density PE (LDPE) with much chain branching, including long side chains, and low crystallinity; and linear low-density polyethylene (LLDPE) with varying degrees of branching, but branches are short, resulting in low crystallinity. HDPE is usually injection- or blow-molded as containers, converted into film for shopping bags, or extruded as sheet or pipe. LDPE and LLDPE are typically made into various packaging films. PE compounds for consumer products require heat stabilizers to protect the polymer from degradation and UV light stabilizers for products used outdoors. Various slip agents, anti-static agents, and other additives have increased PE's usefulness. PE's adaptability has contributed to its increased production over the past decade, even when including the Great Recession period of 2007–2009.

3.2.1.3 End-of-Life

Most PE products are recyclable, at least theoretically. But given the wide range of PE's applications for packaging various contaminating materials, the collection, sorting, and cleaning of post-consumer PE is often too difficult for recycling to be economically viable. HDPE blow-molded as beverage bottles (commonly used for milk), however, is recycled at roughly the same rate as PET beverage bottles in the United States, although the recyclate is reused mainly in non-food bottles and PE pipe.

PE products present many recycling challenges. Not only must LDPE, LLDPE, and HDPE be separated during sorting, the HDPE used for rigid injection-molded containers (like food tubs) has much different melt-flow properties than the blow-molded HDPE in narrow-neck containers (bottles). So these grades may also be separated for creating recycling streams of optimum value. Moreover, unpigmented (natural) HDPE products must be segregated from colored HDPE products in high-value recycling streams. Still, HDPE (along with polypropylene) is the most collected type of non-bottle rigid plastic for recycling [10]. These recyclable rigid products include pallets, tubs, buckets, and household containers. One company, Envision Plastics, based in North Carolina, has realized notable success with its approach to sorting and cleaning HDPE recyclate. The company launched EcoPrime™ resins that can be used in direct-contact food applications. The company runs two manufacturing plants that can recycle between 70-75,000 tons of material per year [11]. In 2017, the company announced a goal to recycle up to 10MM lbs of PE recovered from marine and littoral environments. This ambitious initiative gave birth to "OceanBound Plastic", a recycled PE material that could be used for consumer packaging items such as bottles. Several companies, including Method, became high-profile early adopters. The economics, however, have proved challenging with Envision reportedly unable to find enough demand for the materials they recovered and cleaned [12]. As we will see in later chapters, recycling is not a straightforward proposition, and may not even be the most environmentally optimal choice for end-of-life decisions.

Films made from HDPE, LDPE, and LLDPE are also recyclable, when they are clean enough and can be easily collected. For example, the recycling of blown-film shopping bags is one proposed solution to the plastic bag littering problem discussed in Chapter 1. The used bags' low bulk density, varying material properties, and contamination, however, make recycling difficult. In contrast, a better candidate for recycling is the relatively clean and clear shrink film used for packaging palletized materials. The majority of plastics recycling focuses on rigid containers, partly because they are collected and sorted in the same infrastructure as other recyclable materials, e.g. glass, paper, aluminum. One industry collaborative, Materials Recovery for the Future, has chosen to tackle flexible plastic recycling. In 2016, the group released results from a research program [13] that showed how existing sortation systems could be optimized to capture more flexible plastics. Even though flexible films are measured as being more environmentally friendly than rigids due to their light weight and superior product protection, they are simply not captured or recovered in today's materials recovery facilities (MRFs). This means the end of life scenarios are limited to landfill or incineration, with the light weight benefit negated when films leak into the environment.

PE generally is not biodegradable on its own, but various producers of oxo-biodegradable additives have targeted PE single-use packaging as a key application. These additives break down the polymer chains into shorter pieces over time to make them more likely to be attacked by microbes. There are concerns, however, about how effective the additives truly are for helping turn the polymer completely into carbon dioxide and water, the true end products of hydrocarbon biodegradation (return to Chapter 2 for more discussion). In 2018, The European Commission released a report [14] calling for measures to be taken against oxo-degradable plastics due to a bevy of false claims by some manufacturers. European Bioplastics, an industry organization based in Germany, "strongly welcomed" this report. "While some EU Member States have already set an example and restricted the use of oxo-degradable plastics, including France and Spain, several countries in the Middle-East and Africa are still promoting the use of oxo-degradable plastics or even made their use mandatory [15]." The lack of consensus means that producers of oxo-related additives can continue to sell their goods in markets where the laws allow it. The impact on the marine environment, in particular, will continue to be felt as microplastics increase in prevalence.

In landfills, PE is relatively inert and unreactive (in fact, PE is the base material of landfill sheet liners used to seal off landfills for years). In incineration plants, PE could be thought of as a very waxy hydrocarbon fuel that burns efficiently under the right conditions, releasing mainly CO_2 and water.

3.2.2   Polypropylene (PP)

Another polyolefin, polypropylene, is used for similar applications as PE, though it is generally stiffer and more heat resistant. PP differs from PE in that PP's properties allow it to be used in some durable engineering applications.

3.2.2.1   *Synthesis*

PP is a polymer of propylene, meaning that its polymer chain has a methyl (CH_3) group attached to half of the carbon atoms along the chain (unlike with PE, in which only hydrogen atoms are attached to the chain's carbon atoms). Its production is similar to PE's:

1. Light hydrocarbons are obtained from natural gas or from crude oil refining derivatives (i.e., naphtha, from which propane is more efficiently derived than it is from natural gas).
2. Steam cracking converts propane to propylene.
3. Catalysts are used to polymerize propylene.

The relatively simple processes for creating PP and PE polyolefins means their production costs are very much dependent on the price of oil and natural gas. So extreme and rapid resin price changes are possible when feedstock prices jump, or when natural gas is priced much differently than crude oil, as it was in early 2011. The manufacturing of propylene from sugarcane-derived ethanol is already established by Braskem (Brazil). Nonetheless, the continuing high-volume demand for PP means that for the near future at least, it will mainly be produced via the traditional petrochemical infrastructure as evidenced by the recent completion of Braskem's 1 billion pound facility in La Porte, Texas.

3.2.2.2   Structure and Properties

Polypropylene's relatively simple structure actually allows for much variation. This means that significant property changes result from changes in the polymerization process, depending on where and how the methyl group repeats its relative position along the polymer chain. For example, the methyl groups may all line up on the same side of the polymer chain (in *isotactic* PP), they may alternate one side to the other (*syndiotactic* PP), or they may randomly lie on one or the other sides (*atactic* PP). Each form has different properties. PP's molecular structure and processing also help determine the degree to which its polymer chains crystallize in the bulk material, which also affects its physical properties.

Polypropylene's basic properties can also be changed a great deal by compounding it with fillers and reinforcements. These modifications and its low density and costs have allowed PP to compete with traditional engineering polymers in durable automotive applications. For example, talc filler or glass or other fibers can multiply its modulus (stiffness) and tensile strength values, while increasing its heat-distortion resistance as well. Impact modifiers increase its toughness. PP serves as the base material for what is generally referred to as thermoplastic polyolefin (TPO) – a heavily engineered material that includes a rubbery modifier for low-temperature impact performance plus talc filler for rigidity; it is used for automobile bumper fascia and other large parts.

3.2.2.3   End-of-Life

Polypropylene products are often recyclable, though its recycling is less common than with PE because it is less often used for common, high-volume packaging applications. PP grades are often heavily filled or pigmented, making them harder to sort and reclaim (however, unpigmented and clarified grades of PP are becoming more popular for food packaging, potentially increasing PP's recycling value). Engineered PP film and sheet also create volumes of recyclable material. In landfills, PP, like PE, is inert, and in incineration it produces basic combustion products.

In 2017, PureCycle Technologies announced a breakthrough in PP recycling via a $120MM investment in a new facility in Ironton, Ohio. The technology, developed by scientists at Procter & Gamble, "separates color, odor and any other contaminants from plastic waste feedstock to transform it into ultra-pure recycled polypropylene," according the company website. Since the company was formed, it has attracted significant attention from funders such as the Closed Loop Fund [17] and partners such as Milliken Chemical [18]. With P&G and other major brand owners making public commitments to use more recycled content, the demand for recycled PP is large and growing.

Thus, PP and PE, the common polyolefins, offer recyclability and chemical simplicity. They have rarely been linked with toxicity scares, which have allowed them to develop a public reputation as "friendlier" plastics. They are also relatively soft and flexible, perhaps also helping their image. However, their low densities and low costs are curses as much as blessings, since polyolefin shopping bags and other frequently littered materials have become the target of product bans. The different cost structures between world-scale virgin resin production and regionalized recycling efforts continue to hamper the increased uptake of recycled polyolefins, though several companies continue to invest and find commercial success (see Sidebar on next page).

3.2.3   Polyvinyl Chloride (PVC, or "Vinyl")

No common plastic has created as much antagonism in public discourse as PVC – yet few polymers are used as widely. Whereas a polyolefin can receive a favorable ranking in terms of green chemistry and recyclability in an LCA, PVC more commonly receives failing grades [19, 20].

Yet PVC is effective in durable applications, especially in construction. As an inexpensive polymer partially synthesized from the chlorine atoms in ordinary salt, PVC's most prevalent and dependable uses are "under the radar" – house siding, electrical cable coverings, and window frames. But as discussed in Chapter 1, its use in more personal applications has been reduced over the years, at least partially due to the public's concerns about health issues connected with the phthalate plasticizers used in flexible PVC, and the vinyl chloride and toxins associated with PVC's production and incineration. Though PVC production in the US declined from 2000

Commercial Success in Olefin Recycling

QRS is a large recycler and materials recovery facility (MRF) operator in St. Louis with plants in Louisville and Nashville. The company specializes in post-consumer olefins, primarily PP and PE. After a period of aggressive growth into plastics recovery facilities (PRF), the company was forced to contract during the Great Recession. Believing that diversification would serve as a buffer to future exogenous shocks, company founder, Enrico Siewert, founded Evertrak to manufacture composite railroad ties. His vision – "a world that ships on recycled plastic" – is becoming a reality thanks to the company's ability to take market share from wood products. The life expectancy of a wood railroad tie (one that has possibly been made from imported hardwood and soaked in creosote) is approximately 15 years. A composite tie, by comparison, has a life expectancy of almost 50 years and doesn't require the addition of a toxic substance that can contaminate groundwater and soil. With 25MM rail ties replaced annually in North America, this represents a tremendous opportunity for PCR content without cannibalizing virgin plastic. Beyond ties, QRS is also manufacturing plastic pallets, a market where 400MM units are made every year in the US. Siewert suggested that there is nascent brand equity available for those companies that embrace greater use of PCR resins: participation in the circular economy, energy savings, and a reduction of greenhouse gas emissions.

to 2010, it has seen a resurgence in the past decade, growing 13% during that period [21]. This can be explained in part by a rise in construction activity after major hurricanes along the US Gulf Coast and wildfires in California destroyed large urban areas [22].

3.2.3.1   Synthesis

PVC is a vinyl polymer just as PP actually is, except that instead of PP's methyl group, there is a chlorine atom instead. PVC's production follows the steps of polyethylene's production, plus steps for the addition of the chlorine:

1. Ethane is separated from natural gas or obtained from crude oil refining, while chlorine is produced from the electrolysis of sodium chloride (salt).
2. Steam cracking converts ethane to ethylene.
3. Ethylene and chlorine are combined to form ethylene dichloride.
4. Ethylene dichloride is cracked to produce vinyl chloride monomer and hydrogen chloride.
5. Vinyl chloride monomer is polymerized to become PVC.

Over its decades of its use, PVC's basic composition has slowly raised increasing concerns. Although PVC itself is relatively stable and inert, vinyl chloride monomer is considered toxic and carcinogenic, and must be carefully controlled and monitored in the plant environment. Hydrogen chloride must also be handled properly to prevent its conversion to hydrochloric acid upon release. Recent concerns about chlorinated dioxins in the environment have also called for monitoring whether these compounds are produced and released from PVC production facilities. In 2012, the US Environmental Protection Agency (EPA) proposed stronger emissions standards for all 17 US PVC production facilities, limiting the permissible releases of all the above compounds from process vents, equipment leaks, wastewater, storage vessels, and heat exchangers [23].

PVC's production is also complicated by the intense compounding required to convert inexpensive raw PVC into a processable, useful material. Unplasticized or rigid PVC (commonly used for pipes and construction materials, for example) requires additives that allow it to be extruded or molded. Flexible PVC compounds require high doses of plasticizing additives, commonly added at over 30% by weight to turn PVC into a soft material. Common phthalate-based plasticizers partially exude or leach from the PVC product, and they have been associated with disrupting the endocrine systems of mammals (see Chapter 1 for a discussion of the resulting controversy). This issue is complicated by the number of different phthalates used; some are high-molecular-weight phthalates, which are considered low risk (common types are abbreviated as DINP, DIDP, and DPHP); this supports their continued use in flooring, furniture, and wall covering applications [24]. However, others are low-molecular-weight phthalates (DEHP, DBP, and BBP), which are being restricted and phased out directly by the pressures of regulation, such as Europe's REACH authorization, and now indirectly by pressures from the market itself. Table 3.2 summarizes major phtalate regulations.

The industry is trying to address the phthalate concerns. One approach is to use phthalates that do not migrate to the PVC product's surface, but rather are chemically tied within the polymer. Research has focused on replacing migrating dioctyl phthalates (DOPs) with functionalized versions that bond to the PVC backbone, yet still provide adequate plasticizing properties [26]. Several companies are working to develop plant-based plasticizers using biomass, citric acid or other fatty acids.

Table 3.2 Phthalate regulation in major chemical laws (legal thresholds: 1000mg/kg) [25].

Regulation	Phthalate types
EU RoHS Directive (2011/65/EU)	DIPB, DBP, BBP, DEHP
REACH Annex XVII (EC) No. 1907/2006	DBP, BBP, DEHP, DNOP, DINP, DIDP
CPSIA section 108	DIBP, DBP, BBP, DEHP, DINP, DPENP, DnHP, DCHP
Proposition 65 (California)	DBP, BBP, DEHP, DnHP, DIDP
EU Directive (2005/84/EC)	DBP, BBP, DEHP, DNOP, DINP, DIDP
Legend: di-n-octyl phthalate (DNOP), diisononyl phthalate (DINP), diisodecyl phthalate (DIDP), di-n-hexyl phthalate (DnHP), di-n-pentyl phthalate (DPENP), and dicyclohexyl phthalate (DCHP); EU RoHS: Restriction of Hazardous Substances (European); REACH: Registration, Evaluation, Authorization, and Restriction of Chemicals (European); CPSIA: Consumer Product Safety Improvement Act (United States).	

3.2.3.2   End-of-Life

Methods for recycling unplasticized PVC from construction uses have been developed. But virgin PVC's low price and its additives-intense formulations make recycling hard to justify economically – even with post-industrial materials whose exact composition is known.

Like other traditional polymers, PVC is generally inert in a landfill. However, the incineration of PVC waste produces hydrogen chloride. And, especially when PVC is burned in an uncontrolled way, the chlorine atom in each of PVC's repeating monomer units can become a constituent in organic compounds like dioxins that are extremely persistent in the environment, and which damage human and animal health.

3.2.4   Polystyrene (PS)

Polystyrene is most associated with foam packaging, insulation and plastic eating utensils and cups. It is inexpensive and easily foamed, accounting for the prevalence of expanded polystyrene (EPS) food containers. Contrasting with the single-use nature of these applications is the tremendous energy savings achievable with the use of EPS insulation for keeping food cold and room interiors warm. Still, the production of PS in the United States has decreased by 2% from 2014 to 2018 [27] with similar drops in EU markets. It is being replaced by PET and PP in many applications, particularly in food packaging.

3.2.4.1   Synthesis

Like PP, polystyrene is composed of repeating units of two carbon atoms, one of which has a side unit with a constituent other than hydrogen attached. With PS, the side unit is a large phenyl ring of six carbon atoms and five hydrogens. This structure makes PS a glassy, clear, low-strength polymer usually only suitable for nondurable applications.

Polystyrene's synthesis requires the production of benzene, a carcinogenic material, to form styrene, a suspected carcinogen. Its basic production steps are as follows:

1. Catalytic reforming of hydrocarbons from crude oil, or steam cracking of hydrocarbons from natural gas, is used to create benzene.
2. Reacting benzene and ethylene forms ethylbenzene.
3. Ethylbenzene is dehydrogenated to form styrene.
4. The styrene is polymerized.

Polystyrene is brittle, but the addition of rubber partially grafted to the PS creates tougher grades (i.e., high-impact PS, or HIPS).

PS is converted into EPS for cushioned packaging or thermal insulation using foaming or blowing agents – gases that create the foam's cellular structure. Common foam-blowing agents, especially those used in the past, were blamed for damaging the earth's ozone layer, since they were based on chlorofluorocarbons, most of which have now been phased out. Alternative methods for producing extruded EPS have used relatively benign CO_2 as a blowing agent.

Modifications and additives can improve EPS's properties. For making EPS more efficient as thermal insulation, BASF has developed infrared-energy-absorbing materials that can be integrated with the foam, reducing its thermal conductivity. This allows an EPS panel's thickness to be reduced by 25%, or the density of the foam to be reduced, resulting in a 50% material savings [28].

3.2.4.2   End-of-Life

Waste PS and EPS materials resist biodegradation, and the materials cannot be called "recycling friendly" either, which partially explains some consumers' dislike of single-use EPS coffee cups, for example. Relatively little polystyrene is reclaimed from the municipal recycling stream, though mechanical methods and, more recently, environmentally friendly solvent-based methods are being used to recover PS and EPS. Agilyx, based in Oregon, began as a thermal conversion technology company in 2004, but volatile commodity prices led them to develop a more robust business model whereby they adapted their technology platform to create chemical feedstocks for plastics, breaking polystyrene polymer down to styrene monomer [29]. But EPS foam in particular lacks a large infrastructure for collection and recycling, except for some recent developments for recycling EPS in industrial volumes (although the recycling limitations of EPS should always be weighed against the energy savings provided by EPS as thermal insulation). Dart Container, the world's largest producer of foam cups and containers, has pioneered several attempts to spur EPS recycling, including the "Home for Foam" website that lists locations where the material can be recycled and through the sponsorship of numerous drop-off points. New efforts at promoting circularity for PS include collaboration among resin producers (Trinseo, Berwyn, PA) and converters (Coexpan, Madrid, Spain) to prove the viability of new PS recycling technologies. Still, high-profile battles with consumer groups and city governments have resulted in an increasing number of bans on styrene [30].

As with polyolefins, PS burns efficiently in incineration, though its phenyl ring becomes a source of unwanted organic compounds if there is incomplete combustion.

3.2.5 Polyethylene Terephthalate (PET) and Related Polyesters

PET is a kind of polyester, a general term meaning its components contain ester functional groups. PET somewhat straddles the line between commodity and engineering polymers. Obviously, its highest volume use is in beverage bottles and food containers, and in textile or carpet fibers. Its gas-barrier properties make it the polymer of choice not only for carbonated beverage bottles but also for food condiment bottles (though a layer of polyvinyl alcohol barrier resin is normally required for enhancing barrier properties). PET also has some engineering uses, as does a related polyester – polybutylene terephthalate (PBT), which is used for electrical components.

3.2.5.1 Synthesis

Unlike the commodity polymers mentioned above, PET and most engineering polymers are "condensation polymers," in which two different constituents react to form a long molecule with alternating monomer segments. The repeating unit for PET is created from the reaction of ethylene glycol and terephthalic acid or dimethyl terephthalate (DMT). During the reaction, a polymer chain grows into repeating segments of these combined constituents; process steps generally include:

1. Catalytic reforming of hydrocarbons from crude oil or steam cracking of hydrocarbons from natural gas is used to create *para*-xylene.
2. Ethylene is created from ethane in natural gas or crude oil refining.
3. The *p*-xylene is oxidized to form purified terephthalic acid (PTA) (or subsequently dimethyl terephthalate [DMT]).
4. The ethylene is converted into (mono)ethylene glycol.
5. PTA or DMT and ethylene glycol are reacted stepwise to create PET.

Antimony-based catalysts are used in PET polymerization, and some residual antimony can migrate from the PET after production, but it is normally present at levels well below recommended antimony exposure limits.

The ethylene glycol for PET synthesis can also be made from bio-based ethanol obtained by the fermentation of sugarcane or other biomass. However, unlike bio-PE which is fully bio-based, in which all of the polymer's carbon atoms have a biological source, only 20% of the carbon atoms in bio-ethanol-based PET are from renewable resources. PET that is one hundred percent bio-based can be achieved from the production of bio-based p-xylene, which is then converted into terephthalic acid.

A lower melting polyester similar to PET is polytrimethylene terephthalate (PTT), commercialized as DuPont's Sorona® material. PTT is a partially bio-based polyester option when it is polymerized using plant-based 1,3-propanediol (PDO), instead of the ethylene glycol required for PET (as shown in step 3 above). Standard PDO can be produced using propylene, glycerol, or ethylene oxide derived from petrochemical processing. But biological PDO production is relatively friendly in terms of green chemistry principles and costs [31]. Bio-based PDO is produced as a byproduct of the fermentation of glycerol by a few types of naturally existing microbes. New strains of microorganisms also reportedly allow PDO to be produced directly and cost-effectively using corn-based glucose as a source feedstock. The energy requirements and gas emissions of producing bio-PDO are nearly half of what they are for producing petro-based PDO. Thus the resulting PTT offers bio-based content, while its kinked molecular structure gives it spring-like recovery when deformed, making it useful for carpet or textile fibers.

PEF (polyethylene 2,5-furandicarboxylate) can perhaps be thought of an analog to PET that was first described in patent literature in 1951 [32]. It has enjoyed a renaissance over the past decade thanks to investments in commercialization efforts, primarily led by Netherlands-based Avantium. The main building block for PEF is furandicarboxylic acid (FDCA) which is a bio-based potential substitute for terephthalic acid (PTA). Avantium's efforts have focused on a proprietary catalysis process that converts plant-based sugar (fructose) into chemicals and plastics. The company believes the market potential for PEF could be more than $200bn annually but the pathway to profitability is long and bumpy. Spun out of Royal Dutch Shell in 2000, Avantium has progressed slowly and steadily over the past twenty years, receiving investments from BASF, creating joint ventures, expanding its technology platform, and ultimately going public in 2017. Moving from the pilot phase to commercial production by 2023 is the next hurdle, with an investment of €150MM required to scale-up FDCA and PEF products [33].

Other polyesters are completely bio-based when they are uniquely synthesized from monomers created by fermentation, as in the case of PLA, or polymerized by bacteria themselves, as in the case of PHAs; these are discussed in Section 3.5 below.

3.2.5.2   End-of-Life

Along with HDPE, PET is the most recycled of plastics. Its clarity and relatively high value make recycling economically attractive, depending on collection rates and the quality of the incoming material. There are, however, challenges in PET recycling (and some of these also apply to HDPE recycling); for example:

- Its recycling process consumes a lot of water, as the recycled PET (rPET) flake undergoes intense washing to make it suitable for reuse.
- Because injection/blow-molding and thermoforming require PET resins with different intrinsic viscosities, there is some difficulty in using a single-stream recycling process to create useful rPET material for both thermoformed food trays and bottles. PET bottles may also contain a polyvinyl alcohol layer which acts as a contaminant in some recycling processes.
- Depending on the supply of used PET bottles and the overall health of the economy, the price of recycled PET flake sometimes nearly matches that of virgin bottle PET. This definitely weakens some of the drive for more PET recycling, making it difficult for recycling companies to sustain their businesses.

Despite the large recycling infrastructure, most PET bottles in the United States and about half in Europe end up in landfills or are incinerated or, unfortunately, are littered. In China, PET bottle recycling is more common, with most recycled bottles from North America and Europe being sent there in recent years for reprocessing and reuse. Of course, National Sword has dramatically changed this situation. Still, the past decade shows numerous innovations and successes in PET recycling, with both new entrants and existing industry players making efforts to increase supply and improve quality. Large consumer brands have made very public commitments to using more recycled content helping to drive market signals that should result in investment for more capacity. Companies like rPlanet Earth and Carbonlite, both California-based entities, have responded with significant build-outs of rPET processing facilities. Industry groups such as NAPCOR (National Association for PET Container Resources), APR

(Association of Plastics Recyclers), and Petcore Europe track the recycling rates in North America and Europe. The US PET recycling rate stood at 29.2% in 2017 [34], well below the 58% reported in Europe for the same year [35]. Much could be written to explain the divergence between US and EU collection rates, but suffice to say for now that cultural, economic, geographic, and philosophical differences account for the majority of the explanation.

3.3   Traditional Engineering Thermoplastics

Most petroleum-based engineering polymers could be considered relatively safely ensconced in their key proven applications, where they are used in relatively low volumes compared with commodity thermoplastics. Some biopolymers have been developed that have engineering properties for durable uses (as discussed in Subsection 3.5 below), but generally, any bio-basis in engineering polymers will more likely come from individual bio-based chemical constituents used in their synthesis, rather than in the form of entire polymers uniquely made via biomass conversion processes.

3.3.1   Nylon or Polyamide (PA)

Of all polymers commonly recognized as engineering thermoplastics, the nylon/polyamide family of polymers generally has a lower cost relative to the value of their properties. This makes them extremely important since the mid-1900s for injection-molded mechanical and structural components, as well as for the fibers for stockings, for which nylon was first adapted. Polyamides are easy to mold, but they do tend to absorb water, a factor to be weighed for some applications.

3.3.1.1   *Synthesis*

Forms of nylon/polyamide share the basic characteristic of having an amide (nitrogen-based) group as part of their molecular backbone. Nylons are named according to the number of carbon atoms in their monomers; thus, nylon 6 (polycaprolactam) has a repeating unit containing a chain of six carbon groups, while nylon 6/6 (or PA 6.6) has a monomer built from two different constituents (adipic acid and hexamethylene diamine), each contributing chains of six carbon atoms between the nitrogen groups in the polymer's repeating unit. Nylon 11, nylon 12, and nylon 6/12 are other commercially important forms having higher toughness values than nylon 6 or 6/6.

As a condensation polymer, the traditional synthesis of nylon 6/6, for example, requires the following basic steps which are mostly from the *Handbook of Plastic Materials and Technology [36]*:

1. Cyclohexanone and cyclohexanol is synthesized from cyclohexane (which is created from benzene); these are combined to produce adipic acid. (Adipic acid has also been synthesized from the bacterial conversion of glucose, a greener process [1]).
2. Adipic acid, plus butadiene and acrylonitrile (see the subsection on ABS below), are used to synthesize hexamethylene diamine.
3. A solution of hexamethylene diammonium adipate is created from the products of steps 1a and 1b.
4. The solution is heated to start a polycondensation reaction, producing nylon 6/6.

Nylon 6/10 and 6/12 are synthesized in similar ways, using dicarboxylic acids as feedstocks.

Generally speaking, nylon is a petrochemical-based polymer, but some forms of PA are based on plant materials and produced in significant quantities. Nylon 11's synthesis is interesting from a green chemistry standpoint in that it is completely based on oil from castor beans. This means its production reportedly produces only about 40% as much CO_2 emissions as the production of nylon 6 or 6/6 [37]. Nylon 6/10 (PA 6.10) is partly based on castor oil, thus requiring 20% less fossil fuel than a conventional PA to produce, while creating 50% less greenhouse gas emissions, according to one manufacturer [38]. Nylon 6/10's thermal performance, chemical resistance, gas-barrier properties, and water absorption, are all said to be comparable with traditional nylons' properties.

To enhance their engineering properties, nylons are often compounded with glass fiber reinforcement or mineral fillers.

3.3.1.2   End-of-Life

Post-industrial nylon scrap can be commonly reclaimed, but post-consumer nylon parts require very careful separation methods to be recycled. Along with sorting through the different nylon chemistries, nylon products may have various fiber/filler contents and come in multiple forms, many of which are used, for example, in automotive vehicles. Molded automotive nylon parts are usually labeled with the nylon type and filler/fiber content, but these parts are only starting to be sifted out from the junk-vehicle waste stream as reclaimed materials. Otherwise, automotive nylon ends its life as part of the landfilled "fluff" material that is created when old cars are shredded. Previously mentioned MBA Polymers along with carpet manufacturing companies such as Shaw Industries have proven that nylon can be profitably recycled at scale. In fact, Shaw Industries' Evergreen Nylon Recycling (ENR) facility is the largest of its kind in the world, recovering 25MM lbs of caprolactam and 35MM lbs of calcium carbonate filler annually. The energy savings alone total 450 billion BTU [39].

3.3.2   Acrylonitrile-Butadiene-Styrene (ABS)

ABS is a triple-polymer (terpolymer) system, combining the strength and rigidity of an acrylonitrile component, the rubbery low-temperature toughness of butadiene, and the hardness of styrene. The various components are polymerized together in various ratios, depending on the properties needed in the grade. For decades ABS's useful combinations of properties have allowed it to be used for a wide range of functional products that require mechanical properties and a shiny, attractive surface appearance. Lately, however, compounded polypropylene has been adapted for typical ABS uses.

3.3.2.1   Synthesis

Combining three separate polymer chemistries into one engineered material, ABS synthesis requires multiple production/polymerization reactions. This complexity works against green-chemistry principles. Yet ABS's total production energy is lower than other engineering polymers (Table 3.1), and the petrochemical infrastructure for producing its precursors is mature and developed. The simplified steps are [40]:

1a. Styrene is produced in the steps given above for polystyrene.
1b. Acrylonitrile is produced by reaction of ammonia, oxygen, and propylene (produced via hydrocarbon cracking).
1c. Butadiene is extracted from hydrocarbon streams and polymerized
2.   Styrene and acrylonitrile are combined into a copolymer (abbreviated as SAN).

3.   Styrene and acrylonitrile are also grafted onto polybutadiene to make it compatible with the SAN phase, with which it is blended.

ABS is often compounded with fiber reinforcement for additional strength.

3.3.2.2   End-of-Life

Many obstacles exist to post-consumer ABS recycling. Many ABS applications are physically attached to product structures – such as computer or electronic devices – and cannot be separated by the consumer for recycling. Like other engineering polymers, it does not possess its own resin identification code in the common coding system used in the United States. It is formulated with varying proportions of acrylonitrile, butadiene, and styrene, potentially causing recyclate property inconsistencies. And it is also a relatively low-value polymer.

3.3.3   Polycarbonate (PC)

Polycarbonate's temperature resistance, stiffness, strength, electrical properties, and optical clarity set it apart from transparent polymers such as polystyrene and PET. It has many uses, including optical storage media, lighting, eyeglasses, automotive glazing (windows), and fluid containers.

3.3.3.1   Synthesis

PC is traditionally produced by the reaction of phosgene and bisphenol-A (BPA), which supplies the bulk of the polymer's repeating unit:

1a.   BPA is created by condensation of acetone and phenol, two high-volume petrochemicals (though acetone has also been made using biological fermentation processes).
1b.   Phosgene is produced by combing chlorine and carbon monoxide.
2.   Polycarbonate and sodium chloride are produced by reacting phosgene with BPA treated with sodium hydroxide.

Phosgene has been used in the past as a chemical weapon, and it must be carefully controlled in industrial processes. At least one non-phosgene PC synthesis route has been explored, reacting diphenyl carbonate with BPA, followed by complete PC polymerization in a solid-state process.

Residual bisphenol-A found in PC water bottles and children's products recently has created public controversy (discussed in Chapter 1), after a long history of PC use. Various studies on BPA, mostly funded by government or health agencies, have reported its potential to damage human reproductive system development and other risks. Other studies, mostly funded by chemical and plastics industry interests, continue to refute these conclusions (e.g., [41]); and no single study seems definitive. But the concerns about potential risk have resulted in various nations, regions, and retailers issuing bans on PC products, particularly in fluid-handling products used by children. Potential PC replacements with PC-like properties, such as Eastman Chemical's Tritan™ and PSG's Kostrate® Edge copolyesters, have been commercialized over the past decade.

3.3.3.2   End-of-Life

Despite its high value and wide use in non-food/beverage products such as CDs and DVDs, post-consumer PC is not commonly recycled. Its vague resin identification code number of 7 (other) on products, its wide range of applications, and its relatively low volume of use, all work against improving its potential for post-consumer recycling on a large scale.

With the recent controversies about BPA and resulting bans, PC's history may be on a path that is similar in some ways to PVC's. Its use in food, beverage, or child-contact applications will likely become more limited while it maintains its more entrenched, less controversial uses in data storage, lighting, electronics, and glazing (PC's use in future automotive window glazing may offer a measurable environmental benefit by reducing vehicle weight, thus increasing gasoline mileage). Moreover, the toxic chlorine compound used in its production also works against its green chemistry ranking. Overall, when compared with common packaging plastics and new bioresins, it even has been placed below PVC at the bottom of green design/LCA rankings. This tends to support the concept that the production of more complex engineering thermoplastics may present significant environmental impacts in exchange for their impressive properties.

3.4   Traditional Thermosets and Conventional Composites

This book tends to focus on thermoplastics – materials that can be remelted – rather than thermosets – polymers that cannot be remelted after polymer synthesis. As such, thermosets are sometimes not even spoken of as "plastics" but rather as "composites" (if they contain reinforcing fibers), or as their own specific family of materials (for example, as epoxy, polyurethane, or sheet molding compound (SMC). But many thermosets do similar jobs as thermoplastics, usually in durable applications in which the thermoset components are not expected to be recycled.

3.4.1   Unreinforced Thermosets

Thermosets were the original plastics of the 1800s and early 1900s, though they are now not nearly as widely used as thermoplastics. Thermosetting plastics remain important for specific applications such as electrical industry components or high-heat components, mainly because their polymer chains are chemically locked together and they do not melt.

3.4.1.1   Synthesis

In general, thermosets are formed by reactions of two or more constituents. One of the earliest thermosets, Bakelite, was made by reacting polymer chains into a crosslinked network – a 3-D structure that prevents the molecular chains from flowing against each other. In the case of thermosetting rubbers (also called elastomers), what starts as a viscous liquid ends up as a flexible crosslinked material that does not retain a permanent change of dimension or "set" upon deformation. Other rigid thermosets are not classified as rubbers; they resist yielding to extreme deformation, failing by fracture instead.

Thermosets come from several chemical families, but the following three main polymer groups dominate commercially:

- Epoxies: Used in applications ranging from adhesives and metal food can coatings to industrial molds, epoxies are copolymers synthesized from epoxides and bisphenol-A, also used to produce polycarbonate. Epoxies are popular matrix resins for glass and carbon fiber-reinforced composites (see 3.4.2 below). There are generally no "bio" epoxies, though at least one related bio-thermoset has been developed, composed of epoxidized unsaturated triglyceride resin, cured with polycarbonic anhydrides created via fermentation [42].
- Polyurethanes: PUR is synthesized from a combination of rigid and rubbery segments composed of an isocyanate and a polyol. Depending on its formulation, a PUR can be

molded as flexible foam for lightweight seating cushions or as rigid foam for stiff, lightweight structural panels. Or it can be molded using reaction injection molding (RIM) into solid automotive parts. One drawback to PUR production is that extended isocyanate exposure causes respiratory sensitivity and damage in humans. Compensating for this negative attribute of isocyanate chemistry, green chemistry has been used to create isocyanates based on bio-based feedstocks, or with greener processes that eliminate the use of phosgene. Similarly, PUR polyols have been based on soybean or castor bean oil, adding "bio" content to products such as automotive seating and other cushioning foams [28].

- Unsaturated polyesters: Unlike thermoplastic polyesters such as PET, unsaturated polyesters' molecular chains contain C=C (carbon double) bonds, which, when activated by a catalyst and heat, create crosslinks to form a stable thermoset polymer network. Commercial unsaturated polyesters typically use styrene as its reactive monomer component. The precured material commonly is laid-up by hand with fiber reinforcement in open-mold processes for large parts such as boat hulls; as a result, worker exposure to styrene emissions is a health concern. For example, exposure to styrene at 40 ppm over 15 years reportedly can reduce hearing ability; higher exposures may produce other neurotoxic effects. Thus, in 2011, styrene-using industries considered adopting an occupational exposure limit, limiting worker exposure to styrene to 20 ppm over eight hours [43]. Concerns are also related to styrene being a suspected carcinogen, though recent research suggests that humans lack the enzyme that allows styrene to cause tumors in lungs [44]. Nonetheless, as safeguards, new, low-styrene resins, better ventilation, and closed-mold fabrication processes help reduce worker exposure.

3.4.1.2   End-of-Life

Because they cannot be remelted, thermosets do not fit conventional recycling processes, and their large-scale recycling options are limited. Scrap thermoset material may be ground up and used as a filler to some extent, in the same way that waste rubber products are size-reduced into crumb rubber for various uses. Chemical recycling can be used to reduce the thermosets into usable compounds, such as in the glycolysis of PUR for recovering its polyol building blocks. As with other polymers, thermosets are generally inert within landfills, degrading slowly, if at all. One California-based company, Connora Technologies, developed a low-energy solution called Recyclamine® to recycle epoxy resins. The company was acquired in 2019 by Aditya Birla Chemicals of Thailand [45]. High value materials like carbon fiber can be reclaimed and re-used in a variety of aerospace and automotive applications.

3.4.2   Conventional Composites

In common terminology, "plastic composites" or "fiber reinforced composites" usually refer to a class of structural materials composed of glass fiber reinforcement in either a thermosetting resin matrix (such as epoxy resin) or, more often in recent years, a thermoplastic resin. The materials are often used for structural panels for high-end vehicles and aircraft, though any kind of resin containing reinforcing fibers or fillers might be called a composite. The glass fibers in composites are sometimes in continuous strands or are woven into fabrics to increase directional strength. Or, glass can be added to resin before the molding process simply in the form of short, discontinuous fibers, in lengths of 1–25 mm.

3.4.2.1   Production

Thermoset-matrix composites use the thermosetting resins discussed above. Normally, the unreacted, liquid resin precursors are forced to "wet," or totally impregnate, the glass fiber, which can then be laid up into a molded shape or transferred into a mold. The glass fiber can be difficult or unpleasant to handle in what are often very labor-intensive, hands-on processes. A curing treatment then solidifies the structural composite part.

Epoxy combined with glass fiber is used for prepreg (preimpregnated resin/fiber) structural composites. Glass fibers and unsaturated polyester resin is combined as an uncured bulk material called sheet molding compound (SMC) or bulk molding compound (BMC); it is compression molded and cured into lightweight automotive panels and parts. With polyurethanes, the reactive PUR components, catalysts, and glass can be molded together to produce rigid foams for lightweight structural panels.

3.4.2.2   End-of-Life

Given their crosslinked, non-melting thermoset matrix and high glass content of sometimes well over 50%, standard thermoset composites usually are not fully recyclable. High-heat treatments can remove or burn the resin from the glass fiber, allowing some reuse of the fiber reinforcement. Work has investigated composites using cyanate-based matrix resins that can be chemically decomposed at the end of the product's cycle, outputting undamaged glass and useful chemical derivatives [46]. But otherwise, composites are very useful and durable materials that are destined for landfills.

3.5   Biopolymers: Polymers of Biological Origin

This section will focus on the newest generation of biopolymers, or bioresins, terms that are here defined as polymers synthesized *only* by biological processes using biomass or other naturally renewable, non-fossil-fuel resources as feedstock. These are distinguished from the common polymers discussed above in that although some traditional polymers can be synthesized in conventional processing using chemicals derived from renewable resources, they are usually based on fossil-fuel resources and processes. Thus the term "bio-based" is used generally as a qualifier for polymers or plastics that are composed of substances derived from renewable or biological resources, synthesized using biological or conventional chemical processes. Figure 3.3 below illustrates a simple matrix-based approach to categorizing bio- vs. fossil-based plastics.

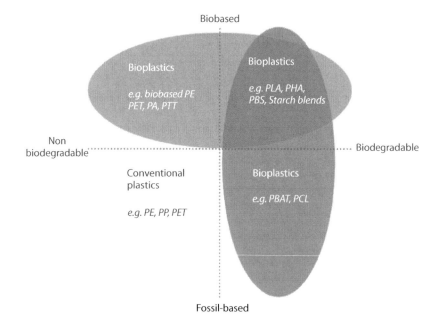

Figure 3.3 Fossil- and Bio-based plastics matrix (Source: European Bioplastics).

The biopolymer sector within the world of plastics presents a range of new terminology to be differentiated and alternative chemistries to be described. This accounts for the lengthier subsections below, compared with those above. The biopolymers focused on here – polylactic acid (PLA), polyhydroxyalkanoates (PHAs), and starch-based polymers – represent the newest generation of bioresins for which significant commercial growth is projected.

Although bioresins currently make up less than one percent of all plastics used, their growth rates are roughly 4x that of commodity thermoplastics. Estimates on the use and production of biopolymers, mainly PLA, PHAs, and starch-based polymers, project growth from 2.11MM tons in 2018 to 2.62MM tons in 2023 [47] (see also Figures 3.4 for total production capacities and 3.5 for discrete bioplastic segmentation). Starch-based bioplastics currently are the highest proportion of biodegradable plastic demand, and they are still growing in use. PLA production rates are expected to reach over one million tons per year within the next decade. And all this growth is projected even with some bio-resins currently costing up to 3x more – much higher than the prices of the traditional commodity plastics they generally compete with (see Table 3.1). Bioresin research and development is a rich area of study, with both start-up ventures and major industry

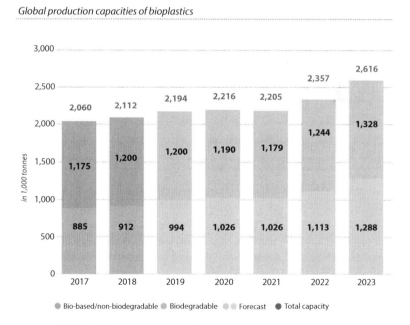

Figure 3.4 Global production capacities of Bioplastics (Source: European Bioplastics, nova-Institute (2018)).

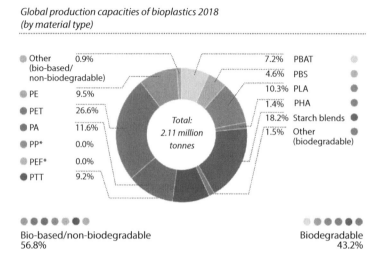

Figure 3.5 Global production capacities of bioplastics by material type (2018) (Source: European Bioplastics, nova-Institute).

players committing resources to new developments. However, a combination of market dynamics, changing regulations, and the slow progress of achieving commercial scale mean that fossil-based resins still dominate.

The current forms of industrial biopolymers come after decades – actually centuries – of human use of biological, natural polymers of various kinds. Biopolymers exist throughout nature in living issues, and biopolymer-based natural materials have been exploited by humans throughout history and prehistory. People have always relied heavily on products made from cellulose and proteins in the form of plant fibers and animal bone, skin, and connective tissues. Starting in the 1800s, natural polymers were modified to give them more useful, controlled properties: natural rubber was refined and crosslinked, and plant cellulose was reacted with nitric acid, resulting in a moldable plastic called celluloid (cellulose nitrate plus camphor). Other bio-based plastics followed, including casein-formaldehyde resin (based on milk protein), and cellophane (dissolved and extruded plant fiber cellulose), which is biodegradable and still used today for food packaging and rayon fiber. More improved commercial bioplastics might have been invented in the 1900s had not the easy availability of petroleum refocused the industry's efforts on developing petrochemical-based polymers – starting with what is commonly considered the first synthetic polymer, the phenol-formaldehyde resin called Bakelite.

Now, after several decades of fossil-fuel-focused effort, environmental concerns have motivated the chemical industry to look again at what polymers and products can be created from the conversion of biomass of all kinds. Conventional chemical processing of biomass can be used to produce various industrial chemicals using gasification, fractionation, and pyrolysis (heating in the absence of oxygen). But these methods can be expensive and do not exploit the abilities of microorganisms, which are incredibly efficient factories for producing chemicals from biomass. Thus, the developments covered below are focused more on biologically produced materials that constitute potentially important commercial directions for the plastics industry. And sections later in this chapter will briefly cover naturally synthesized additives and reinforcements for use in plastics.

3.5.1 Polylactic Acid (PLA)

Polylactic acid or polylactide (PLA) is a kind of polyester that has become a major topic of industry literature and even mainstream media, making it the key, representative biopolymer of current interest. Its range of applications has been expanded to include fiber, film, and molded forms that allow it to compete with PET and commodity polymers in some packaging applications. In food-related applications, such as tea bags, coffee pods, or food serviceware, PLA's degradability under industrial compost conditions is considered useful as it can facilitate delivering more food waste to composters who value those food-borne nutrients. PLA is also used as a common material for 3D printing filament, and in both woven and non-woven applications for hygiene and infusion markets.

3.5.1.1 Synthesis

Polylactic acid is created by the polymerization of lactide, which is created by combining two lactic acid molecules obtained from the fermentation of corn or cane sugars. (Lactic acid can also be found throughout nature in everything from sour milk to fatigued muscles, and it is used in food additives and industrial chemicals.) Commonly based on corn as a raw feedstock source, the basic steps for producing PLA are as follows [48]:

1. Corn is grown with the inputs of sunlight, irrigation water, fertilizers, and insecticides.
2. The corn is harvested and transported to a wet mill facility, where starch is separated from the corn kernels.
3. Using enzymes, the starch is hydrolyzed into sugars.

4. In a fermentation process, the sugars are converted into lactic acid through microbial activity, with calcium hydroxide and sulfuric acid used to control the pH of the production and subsequent purification processes.
5. In a continuous process, the lactic acid is dehydrated into lactide.
6. The lactide is converted into high-molecular-weight PLA, using a solvent-free, catalyzed ring-opening polymerization process.

Even though nature does the work of growing the corn and fermenting the sugar, most everything else in current PLA production is processing- and energy-intensive though on average still less than petrochemical-based plastics like PP, PET, and PS [49]. Energy, typically from fossil fuels, is used for fertilizing, harvesting, transporting, and milling the corn and creating raw feedstock, along with creating the steam and heat needed in PLA production steps. PLA production has become more efficient with scale, with multiple major producers now in operation in the US, Thailand, Japan, and China. In 2019, a peer-reviewed lifecycle analysis (commissioned by Total Corbion) demonstrated that the global warming potential (GWP) of PLA was confirmed to be roughly 75% less than most traditional plastics [50].

The carbon contained in the biopolymer is drawn from the atmosphere via short-term plant growth, rather than from old carbon stored underground millions of years ago. And, unlike the fermentation process used to produce ethanol fuel and feedstock, lactic acid fermentation can almost completely convert glucose into PLA, with a theoretical yield of about 80% after polymerization (compared with 31% for ethanol-based PE) [9]. Thus PLA incorporates atmospheric carbon efficiently, theoretically returning the carbon to the atmosphere after PLA biodegradation, allowing a new carbon cycle to begin.

However, other environmental impacts result from PLA production that make it look less attractive than more straightforward polyolefin production. Eutrophication caused by runoff from large-scale, fertilizer-based agriculture damages the quality of water bodies, and this is something which contributed to PLA's and other starch-based materials lower rankings in LCA studies [51]. Ozone depletion and acidification are other areas where biopolymer production has greater impacts than traditional polymers. Otherwise, PLA's production methods score well in terms of green chemistry principles, given its use of a biochemical process that adapts processes from nature, rather than relying on solvents and precursors.

3.5.1.2   Structures and Properties

Lactic acid's and PLA's molecular structures have a chiral, or "handed" character in the configuration of their atom groups, just as in the way the fingers and thumb of our left and right hands mirror each other in their relative placement. The PLA structure has two mirror-image stereoisomers, depending on which enzymes are used to create the material. The two forms are labeled "D" (as in poly-D-lactide, PDLA) and "L" (poly-L-lactide, PLLA). PLA's properties are determined by the ratio of these two forms present in the polymer chain. An even (racemic) ratio of D and L forms, results in a relatively weak, amorphous material [52]. But PLLA containing only some D units, for example, results in a more useful, slow-crystallizing material. Nearly pure PLLA or PDLA homopolymer can crystallize even more, reaching a maximum melting point of 180 °C [9, 53].

Pure commercial PLLA has properties that make it similar to and a potential replacement for PS, PET, and polyolefins in some uses. In its semicrystalline form, it has good strength, toughness, and adequate heat resistance for room temperature uses. But commercial PLA is slow to crystallize in conventional conversion processes, meaning PLA products are often mostly amorphous, which tends to limit it to applications that are not heated above room temperature. So even though it is a relatively rigid material at room temperature, it has a limiting glass-transition (softening) temperature of about 50–80 °C [54].

Increased crystallization of the PLA polymer improves its properties. Some useful stress-induced crystallization does occur in PLA during thermoforming, as with PET. And crystallization can be enhanced by polymer orientation induced during the extrusion of PLA films and fiber. Additional annealing (heat-aging) treatments also induce crystallization, but they add costs, making PLA less competitive with other polymers. Alternatively, though it seems counterintuitive, blending in some pure PDLA polymer with PLLA has been found to enhance PLLA crystallization, while also increasing the melting temperature and heat-deflection temperature.

To encourage its crystallization, many efforts have focused on developing special additives for PLA. Talc, polyethylene glycol, and microcrystalline cellulose fillers have all been tested in attempts to improve both processing and performance of PLA.

Additives technologies like these will be critical for driving PLA's commercial growth and widening its application range, though ideally they too should be biodegradable or at least bio-based. Gas-barrier polymer layers or additives in PLA products are required to enhance PLA's limited barrier properties. Blends of PLA with other polymers also will drive applications growth (see Subsection 3.5.6).

Beyond new Ingeo®-based product introductions from NatureWorks, several leading players in the biopolymer ecosystem have announced developments in additive masterbatches such as impact modifiers for transparency, slip and antiblock agents for coefficient of friction changes, and barrier films for enhanced shelf life applications. Sukano (Duncan, SC) has done work on a variety of additives including a transparent modifier for use in thermoformed packaging. This nucleating masterbatch forms many small nucleation sites which increases the speed of crystallization. A melt strength enhancer improves IV and mechanical properties while retaining transparency. Using LDR of 1-3% is shown to reduce brittleness without affecting temperature resistance. For injection molding grades of PLA, Sukano S687-D, an opaque impact modifier, improves elasticity and toughness for high-stress applications when used at 10-30% dosage rates [55].

3.5.1.3 End-of-Life

PLA products are designed to be fully biodegradable in industrial composting operations. Biodegrading PLA requires controlled composting conditions and temperatures over 60°C [56]. PLA will not degrade in common outdoor environments, and thus PLA products have the potential to become litter. In the right composting conditions, PLA undergoes a two-step degradation process of disintegration then bio-degradation.

PLA can also be incinerated and converted into carbon dioxide, theoretically a carbon-neutral event because the PLA's carbon was previously atmospheric carbon. This is unlike with traditional plastics, which bring fossilized carbon up to the surface ecosystem – creating a net carbon gain to the atmosphere if the plastics are incinerated. However, because of the fossil fuel energy used for producing PLA, it still has a small net carbon-positive effect, overall adding new carbon to the atmosphere, no matter whether it is incinerated or landfilled. But this effect is still less than one-third of the CO_2 impact of producing virgin PET resin, which produces over 3 kg of CO_2 per kg of PET produced [56].

As with PET, PLA's net carbon impact can be offset by effective recycling. Unfortunately, PLA's temperature and water sensitivity makes it incompatible with conventional mechanical/melt recycling methods used for traditional commodity thermoplastics. It is even considered by some as being a contaminant in plastic recycling streams, because clear, rigid PLA packaging is visually difficult to distinguish from similar recyclable polystyrene or PET packaging. Advances in optical sorting technologies such as near-infrared systems (NIR) are helping to address these challenges, however. As discussed in Chapter 2, bans of PLA packaging have been proposed until its recycling issues are resolved.

3.5.2 Polyhydroxyalkanoates (PHAs): PHB and Related Copolymers

The term PHA comprises a family of biopolyesters created via fermentation processes within different bacteria found in various natural environments. These same kinds of bacteria also degrade and consume PHA polymer products after they are disposed of.

PHAs have been studied for many years as natural polymers that are stored as food-energy sources within bacterial cells. Various bacteria and enzymes produce different PHAs with various structures; over 150 PHA monomer structures have been reported [57]. Over the past couple of decades, research and development has focused on exploiting some of these materials as hydrophobic, bio-degradable, gas-impermeable thermoplastics. This century, activity by Proctor & Gamble and Metabolix, Inc., supported by large agricultural conglomerate Archer Daniels Midland Company, has resulted in a number of commercialized PHA grades branded as Mirel™ biopolymers†. These materials have been tailored for plastics conversion processes for creating packaging and other (mainly) non-engineering plastics applications, though achieving commercial scale has proven expensive and difficult. Danimer Scientific (Bainbridge, GA) recently completed work on a production facility in Kentucky to manufacture its line of Nodax™ PHA resins. In November 2019, the company signed an exclusive agreement with US-based Genpak to provide PHA for takeout food containers [58].

PHAs of key commercial interest are related to poly (β-hydroxybutyrate), also referred to as poly(3-hydroxybutyrate) or poly(hydroxybutanoic acid) – all commonly abbreviated as P3HB or just PHB. This form of PHA is a brittle, crystalline polymer with low elongation properties and limited usefulness. Tougher, lower-melting biodegradable polymers and copolymers have improved PHAs' commercial potential for competing with traditional commodity plastics in many applications; two of these key materials are:

---

† In 2017, Metabolix became Yield10 Biosciences, an agricultural biotechnology company. The Mirel tradename is carried by Telles, an Archer Daniels Midland company that was the original commercialization partner of Metabolix from 2006 to 2012.

- PHBV, a copolymer of P3HB and another PHA, poly(3-hydroxyvalerate) (P3HV or PHV). PHBV can be synthesized as a block copolymer with durable properties. Increasing the PHV content increases ductility and lowers stiffness, while lowering the copolymer's melting and glass transition temperatures. Some PHBV grades have properties similar to those of polypropylene [54].
- PBAT (poly[butylene adipate-co-terephthalate]), a tough, biodegradable copolymer that can be used as a toughening agent in slowly crystallizing PHB copolymers, which can become brittle during aging [58, 59].

PHAs are important biocompatible materials used in medical implants, tissue engineering, or drug delivery. They have also been converted into several kinds of plastic products, such as shopping bags, carpeting, special garments, and shampoo bottles and other disposable personal hygiene products. But the monetary cost of PHA production (see Table 3.1) limits their widespread use, at least for now.

3.5.2.1   Synthesis

The production of PHA polymers is more like a harvesting operation than a chemical synthesis process. PHA production methods can vary, and their details are typically closely-held trade secrets; in general they use these basic steps (mostly summarized from *Biochemical Engineering Journal [60]*, except when otherwise noted):

1. Selected or modified (metabolically engineered) bacterial strains (such as *E. coli*) are cultivated using glucose or other materials as a food source. Via fermentation, the bacteria produce PHA and store it as granular inclusions within their cell bodies. Strains are selected based on their yield in producing the desired polyhydroxyalkanoate, such as PHB; some strains of bacteria have cells composed of up to 29% PHA [61].
2. The bacteria are exposed to heat, freezing, alkaline treatments, or salt-solution treatments to weaken their cell structures.
3. Various harsh extraction treatments further disrupt the cell structures of the bacteria, liberating their PHA granules.

Many methods have been employed, some of which damage the PHA or have other drawbacks. Solvents can be used for extraction, though the solvents may include chlorinated hydrocarbons that are unwanted in a "green" production plant. "Digestion methods" use surfactants that disrupt the cell's membrane, releasing PHA, but effective digestion methods rely on the use of bleach and chloroform. Mechanical bead milling, high-pressure homogenization, and other extraction methods have also been cited as options. The harvesting of the polymer creates significant biomass waste from the cell structures.

4. Controlled purification separates useful PHA from impurities. Heated hydrogen peroxide plus enzymes or chelating agents may be used; ozone treatments have also reportedly been developed.

To create properties more similar to those of PP or ABS, various modifications can be made to raw PHAs such as epoxidation, grafting, block copolymerization, and crosslinking [57]. Post-fabrication annealing/aging of copolymer PHB has also been found to significantly increase its tensile modulus while decreasing tensile elongation at break and tensile toughness. These changes are said to reach plateaus after about 10 days at room temperature [61].

3.5.2.2   End-of-Life

Polyhydroxyalkanoates' biodegradability provides similar potential benefits as the ones cited above for PLA. The key difference is that PHAs are much more biodegradable in common natural environments, outside of controlled composting conditions. Various PHA/PHB polymer and copolymer grades are naturally biodegradable in soil or even marine environments [61]. Marine biodegradation is impeded by cold water temperatures and limited microbial activity. But significant marine biodegradability of PHBV film in natural seawater has been shown to occur in testing, as per ASTM D6691–09 test methods [62]. Here, PHBV film reportedly showed over 75% biodegradation (mineralization) within 35 days. Such biodegradability could open up opportunities for PHA packaging film in coastal communities, or even for shipboard applications in which marine-compostable wastes are regularly thrown overboard (assuming the PHA poses no danger for the marine life that eats the waste).

3.5.3 Starch-Based Polymers

Analogous to PHA/PHB in bacteria, starch is the material stored as polysaccharides in plant cells, providing energy-rich carbohydrate foodstuffs. Starch's two component structures are amylose, a linear polymer, and amylopectin, a branched polymer; both have high molecular weights. These components (present in varying ratios depending on the starch source) combine to form a material that is about 20–45% crystalline [53].

Starch is partially or fully soluble in water, or else is swollen by water, depending on temperature, shear, and pressure. Thus, starch must be modified during processing to become usable thermoplastic starch (TPS), allowing plastics-like packaging applications. Without the addition of plasticizers or blending with other polymers (see Subsection 3.5.6), starch's moisture sensitivity, brittleness, and other disfavored properties would strictly limit its applications, preventing it from being used for blown-film trash bags or thermoformed products, for example [63, 64].

3.5.3.1 Synthesis

Obvious concentrated sources of starch include corn, wheat, and potatoes. The natural starch granules must first be "gelatinized" with heat and moisture and extruder processing [64]. Because starch's melting temperature (220–245 °C) is close to the temperature at which it starts to degrade, plasticizing agents are added to make the starch more processable; these agents include polyols (such as glycerol), amides, and citric acid.

3.5.3.2 End-of-Life

TPS is biodegradable to the point that the shelf-lives and use-environments of TPS products are limited, as is its recycling.

3.5.4 Protein-Based Polymers

Ever since Henry Ford slammed an axe into a car's deck-lid panel made from an experimental soy-based resin (without resulting damage), researchers have looked to turn agricultural and natural proteins into materials that can be widely commercialized. Soy protein is an abundant source for industrial uses, though only recently has serious work been done to commercialize soy protein plastic for biodegradable packaging and durable products.

3.5.4.1   Synthesis

Soy is an obvious candidate feedstock source for protein-based polymers, given the degree to which it is cultivated and its high protein content. Proteins are composed of amino acids, which contain functional groups bonded into polymer chains with various structures and properties.

Soy proteins have a strong tendency to absorb water, severely reducing their strength and usefulness. Work has been done in blending proteins with other biopolymers, and in using up to 10% anhydride compatibilizing agents such as maleic anhydride and phthalic anhydride. These blends reportedly dramatically lower water absorption, helping the protein polymer to maintain its strength [65].

3.5.4.2   End-of-Life

Protein polymers are biodegradable and hydrophilic to the point that the recycling of protein products is limited or impossible.

3.5.5   Algae-Based Polymers

Algae grown in large-scale controlled environments have the potential of producing oils that can be converted into useful polymer feedstock such as hydrocarbons and ethanol – essentially making an algae colony a solar-powered factory for producing renewable raw materials. The high-density cultivation of algae for producing chemicals and fuels would eliminate concerns about using food crops for purposes other than food. As a source of ethanol, algae are potential alternative sources for producing bio-polyethylene and other commodity polymers. Commercial attempts to introduce algae-based plastics to the market have not been successful, though at least one US-based company, Algix, has developed a method to convert algae blooms to plastic pellets which are subsequently foamed for use in footwear soles [66].

3.5.5.1   Synthesis

Algae grow fast, compared with corn or other crops, with growth and harvesting taking place in cycles of about one week. Algae can also be grown in seawater or in efficient man-made ponds or photobioreactors, requiring much less land than corn for producing the same amount of biofeedstock.

3.5.5.2   End-of-Life

Algae-based polymers designed for commercial use would have similar biodegradability, recycling, and disposal issues as competing bio-based polymers of the same polymer family.

3.5.6   Blends of Biopolymers

Traditional polymers are often blended to optimize their desired properties. For biopolymers, the development of blends may be even more important, because common biopolymers have properties such as brittleness or slow biodegradability that limit their uses. One biopolymer's properties can often be improved by blending it with another. For example, the bio-copolyester PBAT is blended with PLA to capitalize on one's ductility and the other's stiffness [67].

Thermoplastic starch's properties have been improved with the addition of PLA, and vice versa. Increasing levels of TPS in PLA reduce the blend's tensile strength and modulus, while increasing its elongation and biodegradability. However, the two polymer types are generally incompatible, requiring coupling agents such as maleic anhydride to improve their interaction.

Blended-in TPS can also be effective as a cost-cutting component when it is blended through melt-processing with more expensive polymers, such as PLA. TPS has also been blended with the high-cost biodegradable synthetic polymer polycaprolactone without severely reducing PCL's mechanical properties, potentially expanding PCL's use [68].

Starch has also been used to increase the biodegradability of bio- and non-biopolymers, whether used as a filler or melt blended. Biodegradability claims, sometimes hotly debated, have been made about blends of traditional polymers and starch. But starch can at least enhance a biopolymer's already inherent biodegradability. With PBAT, up to 40% melt-blended TPS increases the material's ASTM D6400 compostability while retaining its properties, allowing the blend's use for bags, cutlery, trays, and plates, according to manufacturer Teknor Apex Company [69].

Biopolymers can also be blended with traditional polymers, thus enhancing their biocontent while enhancing (or only minimally harming) certain properties. For example, PLA and PC have been blended to enable PLA's use in durable electronics applications; PLA has also been blended with ABS and poly(methyl methacrylate). However, the only partial biodegradability of these blends creates concerns that small, polluting pieces of petroleum-based polymer would remain behind after initial degradation.

Bioplastics continue to be used in a growing array of applications fueling more investment in capacity and discrete developments. Packaging still remains the largest area as chemists and engineers pursue a balance between consumer demand and environmental exigencies. Europe has become the central hub for research and development of bioplastic while Asian countries account for the majority of production sites.

3.6   Additives and Fillers: Conventional and Bio-Based

For a producer of a plastic to be able to claim that a material is 100% bio-based, both the entire polymer and all its additives must be made from material of renewable or natural origin. Many common additives used in plastics are already based on natural-occurring compounds, such as fatty acids. Many fillers and reinforcing fibers discussed below are also of natural origin and have low environmental impacts.

3.6.1   Common Additives

Additives are materials added in small percentages to a polymer to protect it from oxidation by heat or UV light, to improve its melt flow or flexibility, to encourage its crystallization, to toughen it against impacts, to lubricate its surface, to reduce its electrical static build-up, or to color it. Usually, the less complicated the technical expectations are for an additive, the more likely it is that the additive can be based on a relatively "green" natural source. However, new bioresins may require the development and use of more complex naturally occurring additives in order to reach the goals their manufacturers have for them.

Some key, high-volume types of additives are summarized below, with an emphasis on the "green" character of their chemistries:

Antioxidants and heat stabilizers in the past had commonly been based on the heavy metals cadmium and lead, but today are being replaced by metal carboxylates based on the "friendlier" elements of calcium, magnesium, and zinc. Stabilizers are important in PVC processing, and antioxidants are used to protect polyolefins and other polymers from oxidation.

All these additives are commonly petrochemical-based, but antioxidants based on naturally occurring Vitamin E, as well as cashew nut shell liquid and other plant extracts, have been developed, though not yet with any commercial scale or success for the plastics industry.

Colorants for plastics have typically included special inorganic or organic compounds that are not often bio-based. Manufacturers have released colorants for plastics that are said to meet standards for biocontent and composting as per ASTM D6866 and D6400 [70]. These pigments use non-petrochemical-based carrier materials that allow them to be dispersed into the polymer.

Plasticizers for flexible PVC – especially the conventional phthalate-based materials that have raised health concerns – can have a major influence on the green chemistry/LCA rating of a formulation when they are used at high loadings. Non-phthalate plasticizers have been commercialized by Dow Chemical and BASF; some of which are reportedly made from renewable feedstock, thereby letting PVC users claim some bio-based content in their products. The plasticizer called DINCH (1,2-cyclohexane dicarboxylic acid diisononyl ester) has been proposed as an option with a lower life-cycle impact [71]. Isoboride-based plasticizers are another greener option. Other bio-plasticizers are based on citrates, vegetable oils such as palm oil, or even depolymerized natural rubber [72].

Impact modifiers are particularly needed for brittle PLA. Most PLA modifiers are acceptable in food-use products, the largest current application area from PLA, and do not interfere with its biodegradability. Some modifiers also improve processing limitations (like poor melt strength and metal release), though they may cloud the transparent clarity of the PLA.

Flame retardants in plastics have had a poor environmental reputation because the cheapest and most effective ones use halogen-based chemistries such as polybrominated diphenyl ethers like decabromodiphenyl ether (decaBDE). Alternatively, non-halogen-containing, mineral-based flame retardants such as aluminum trihydrate (ATH) and magnesium (di) hydroxide (MDH) are becoming more commonly used as greener alternatives. But these fillers affect the mechanical properties of the plastic when they are incorporated at their effective, flame-retarding loadings. Phosphorous-based flame retardant compounds are another increasingly popular halogen-free approach, though these may have some questionable environmental impacts. There are polymer based FR's on the market now, such as the Dow Bluedge™ product.

Lubricants, slip agents, and anti-blocking agents allow the plastic melt to be more easily processed, or prevent surface sticking in thin-film plastics. Lubricants for common commodity plastics are often simple hydrocarbon waxes, though other lubricants can have more complicated chemistries. Slip agents allow plastic film to slide against itself or metal machinery, and they are commonly based on plant or animal fatty acid amides that are generally approved by regulatory agencies for food packaging use or human contact.

Antiblocking agents are typically small, inert filler particles that prevent stacked films from bonding together.

Antistatic agents prevent static build-up by dissipating electrical charge at the plastic's surface. Fatty acid esters or amines are used in common applications, and are approved for use in food packaging.

3.6.2   Fillers

Calcium carbonate ($CaCO_3$) is the most common inexpensive plastics filler that is literally a "filler" – its main value is to displace polymer in a product's composition. It essentially reduces the amount of resin used in a product, which is especially useful when resin prices are high. Fine particles of $CaCO_3$, loaded at 10–60%, can also increase a plastic's stiffness, hardness, and dimensional stability. Environmentally, the mineral is naturally occurring (from limestone), chemically simple, and commonly acceptable in food-contact packaging. And because it is simple relative amount of processing it undergoes in its production is low, even when compared with the relatively simple polyethylene resins it is commonly used in. Being a mineral, calcium carbonate does not degrade much from the heat of reprocessing as polymers do, and it even lowers the processing energy of polyolefin conversion process. Thus it comes close to being an "environmental footprint-reducing" additive.

Talc and mica are plentiful, inert, and naturally occurring metal-hydrate silicate mineral fillers. They are used in engineering applications in polypropylene, adding properties to the low-cost polymer, such as increased modulus and heat-deflection temperature. The fillers' mechanical-property effects are produced by their plate-like particles' high aspect ratios (the ratio of a particle's major dimension [length] to its minor dimension [thickness]). Talc in particular allows PP to compete in plastics applications that typically require more complex and more expensive engineering resins. Thus, talc and mica could be seen as green-enabling technologies. They allow reduced use of relatively complex engineering polymers such as nylon, while expanding the use of simpler ones like PP, which have less complicated chemical synthesis processes and better recycling potential.

Wollastonite is a less often used mineral filler with needle-like particles (which are not hazardous like asbestos). At 10–20% loadings, wollastonite can increase a polyolefin's tensile and flexural strength, while offering higher dimensional stability and less mold shrinkage than talc or $CaCO_3$.

3.6.3   Fiber Reinforcement

3.6.3.1   *Glass and Carbon Fiber*

The reinforcement of plastics using integrated glass fibers has been done for many years, and has allowed plastics to make the leap from low-value, low-performing applications to high-value, lightweight engineering applications. Glass fiber-reinforced plastics are common in automotive uses, while carbon-fiber plastic composites are growing in importance in the aircraft industry as a way of making planes lighter. Glass fiber is relatively cheap for the properties it creates; carbon fiber is expensive, though it has mechanical properties close to those of steel.

In environmental terms, the use of glass fiber-reinforced plastics has created a contradiction related to their main use in the transportation sector. One the one hand, reinforced plastic parts make vehicles lighter, saving fuel during their use; on the other hand, these plastics are very difficult or impossible to recycle, especially if the matrix resin is a thermoset. Life-cycle assessments have tried to take into account both of these impacts, and any LCA would have to weigh the importance of greenhouse-gas/fuel savings from "lightweighting" vehicles using these materials against the environmental impacts of fiber and plastic production and disposal.

Carbon fiber is a stronger, lighter alternative to glass fiber, though currently it is too expensive for all but the most specialized, high-end transportation and sports applications. Its mechanical properties are of a higher order of magnitude than any plastics materials; the modulus of carbon fiber is about 50 GPa, its tensile strength is around 3,000 MPa, and its linear thermal expansion is near zero. It also provides electromagnetic interference and radio frequency shielding for the composites it is used in [71].

Both carbon and glass fiber have sustainability problems related to their recyclability. About 30% (24,000 metric tons) of all produced carbon fiber ends up in landfills [73] and total glass-fiber composite scrap is certainly much higher, given that it is by far the predominant fiber-reinforced composite material. Recovering the fiber by separating it from its polymer matrix material is difficult; efforts to separate fibers mechanically, chemically, or thermally tend to damage fibers by introducing stresses and surface defects or reducing average fiber length. In 2015, a non-profit called Composite Recycling Technology Center (CRTC), was launched to address the challenges with recycling carbon fiber. Since inception, the group has developed a park bench made from recycled aerospace-grade carbon fiber. With a price tag of $4500 [74], it remains to be seen how commercially viable the technology will become, though the growth of wind-energy and aerospace industries in particular will drive demand for innovation. Other notes about glass fiber's environmental impacts are integrated below, alongside information about the natural plant fibers that are being considered to replace glass in plastics.

3.6.3.2   Natural Fiber Reinforcement

Fibrous plants and wood have always been useful renewable sources of structural materials. In the world of plastics, cellophane and rayon are well known forms of chemically modified cellulose from plant fibers, but relatively raw plant fibers themselves are becoming increasingly important reinforcing agents. Used in traditional plastics, sometimes as an alternative to glass fiber, plant fibers add bio-based content to the composite material, while also enhancing its engineering properties. And in biopolymers, plant fibers allow a bioplastic to stay "green" in character while expanding its potential uses. Plant fiber reinforcement also has the potential to reduce the energy footprint of engineering plastics in many situations.

Fibers from various fast growing plants can be used in plastics; some useful plant fibers are otherwise treated as waste materials from agricultural production. Moreover, these plant fiber sources are spread throughout the world, allowing developing countries a chance to participate in the global market for industrial materials.

Plant fibers are generally composed of cellulose, with varying amounts of lignin (up to 50% in wood). Both materials are natural polymers. The use of plant-fiber reinforcement is somewhat complicated by its low heat resistance and by the effects of water on its properties – two non-issues with glass fibers. However, plant fibers' sensitivity to water and environmental exposure is an indicator of their biodegradability.

Having properties that start to degrade above 170 °C, plant fibers often cannot be processed in many of the high-pressure/high-temperature processes used in molding glass-reinforced plastics. Rather, plant fiber composites are often integrated with a polymer as laminated fiber fabric mats or sheets, molded in low-pressure, compression-molding processes. And they are generally weaker and less stiff than glass fibers, with their fiber sizes and shapes and properties showing more variation. Moreover, hydrophilic plant fibers do not naturally bond well with hydrophobic polymers, and thus fiber surface treatments or coupling agents are used to improve composite properties.

Therefore, plant fibers are typically not "drop-in" replacements for glass in engineering plastic applications that require fiber reinforcement. But when they can be used, plant fibers provide weight saving opportunities and other environmental benefits. Low-density plant fibers thus can make cheaper, greener materials like polypropylene or biopolymers more competitive for applications for which unfilled engineering plastics would otherwise be chosen. In life-cycle comparisons with glass fiber, plant fibers also have several advantages. Generally, they require less energy to produce, and they produce less air pollution, creating lower emissions (CO_2 and otherwise). Their production can release nitrates and phosphates into water systems, but only when fertilizers are used for growing plants dedicated to fiber production [75]. At end of life, they are biodegradable, though they are sealed within a polymer that might not be. Their recyclability is challenged because of the damaging effects of repeated heating when the polymer/fiber composite is reprocessed.

Various natural fibers are already cultivated for industrial purposes in high volumes. High-quality plant fibers cost about the same as or less than glass fiber per pound, though glass provides higher mechanical properties per unit cost. Thus the use of natural fibers may require using a more expensive polymer resin with higher properties than would be required than when using glass-fiber reinforcement, to compensate for the lower plant fiber properties. However, when considering the much lower densities of plant fibers (1.3–1.5 vs. 2.5 for glass), and their strength and modulus values per unit density, they appear more competitive. Therefore, a version of a reinforced polypropylene part incorporating plant fiber may weigh less than an equivalent one that uses glass-fiber reinforcement. Natural fibers' value is also enhanced in comparison with glass fiber in that glass is abrasive, making it unpleasant to handle in a plant and damaging to mold and tooling surfaces [76]. The natural fibers of interest summarized below (except for wood and coir) are composed of roughly 0–20% lignin, which is a nonlinear natural polymer in plants that acts as a binder of plant fiber components. The remaining components of these fibers is 60–80% cellulose (a strength providing linear polymer), plus varying degrees of wax, moisture, and hemicellulose (an amorphous, short-chained, branched polysaccharide). The examples of plastics applications mentioned below illustrate the kinds of problems that are being addressed in making natural fibers optimally practical and sustainable in their use.

Jute is second only to cotton in terms of global fiber production with India and Bangladesh as the largest producers. In one plastic process, continuous jute fiber fabric mats (laminated with PET sheet and treated with shellac, another natural polymer) were used in PLA insert-injection molding to create a stiff composite [77].

Kenaf fibers are taken from the middle of the fast-growing plant's stalk; they display high tensile strength (261 MPa) and modulus (20 GPa) [71].

Hemp stem (bast) fibers are often converted into twine and nonwoven mats. These forms have been investigated for use in automotive sheet molding composite material instead of glass (or in a combination of 45% hemp and 5% glass fiber), but like other natural fibers, hemp's use is complicated by its water absorption.

Flax bast fiber is commonly used for textiles, or in coarse form, for twines and canvas. Fibers are extracted after a period of partial natural decay of the flax plant stalks, which makes their useful "technical fibers" in the plant stem easier to separate [54]. Flax fibers as woven fabrics have been compression-molded with PP and used in twin-screw extruders via solid-state shear pulverization (SSSP) with improvements in stiffness and strength [78].

Wheat straw fiber has been developed as a reinforcing filler for injection-molded, PP-based automotive parts, which are currently serving in actual vehicle applications [79]. Wheat straw comes from the wheat food crop's plant stems which would otherwise be treated as agricultural waste. Waste fibers from the agave plant and sunflower seed husks have similar potential.

Bamboo contains very strong fibers, and it has been commercialized as a reinforcement for an injection-molding grade of PP [71]. However, bamboo has low compatibility with thermoplastics; this low interfacial strength has been addressed with additions of "micro-fibrillated" cellulose [80]. Bamboo fibers have also been made more compatible with PLA by pretreatment with alkali, which helps remove lignin from the fiber's surface, or by using silane coupling agents.

Sisal is a leaf-extracted fiber used for various industrial purposes, such as floor coverings.

Agave fiber is related to sisal, and comes from the plant used to make tequila and sweetener. Otherwise treated as a waste fiber, it has been used in PP at 55% loading, reportedly increasing PP's tensile modulus when PP is grafted with maleic anhydride coupling agent [71].

Abaca (banana) fiber is among the cheapest fibers available for use in plastics, costing well under $1 US per kilogram [75]. Abaca fiber, which comes from waste banana plant husks, has been used in both interior and exterior automotive applications by Mercedes Benz [81].

Cotton seed fiber is composed of over 80% cellulose, higher than all other fibers mentioned above, with no lignin. When used to reinforce thermoplastic starch, cotton fiber is said to also promote the biodegradation of the TPS in soil [82].

Coir (coconut) fibers are high-lignin materials from the protective husks of immature or mature coconuts. Coir fibers are relatively resistant to fungal and microbial degradation, even in wet environments; this means that piles of coconut husks can be accumulated over time, with the fiber remaining usable during storage and shipping. Coir/plastic composites also reportedly produce little odor when used in molded parts. Nonwoven coir fibers have been mixed with polypropylene fibers to create lightweight 50% coir/50% PP compression-molded automotive panels [83].

Wood fiber normally contains only up to 50% cellulose, with a higher percentage of lignin than other natural fibers. Wood fiber is abundant, especially in the form of sawdust or other post-industrial wood, and it has become useful in wood-plastic composites used for decking and construction materials. These materials also often incorporate recycled commodity resins, making them a longer lasting, "green" alternative to real wood (see Chapter 4 for examples). Cellulose fibers derived from wood pulp (such as newspaper fiber) have ribbon-like shapes, facilitating load transfer to and from the polymer. Stripped of lignin and hemicellulose, the cellulose in pulp fiber is a high-aspect-ratio fiber that costs less than glass fiber. It has good chemical resistance and heat resistance up to 260 °C [71].

As mentioned above, one problem with natural fibers as reinforcements is that they resist bonding with most polymer matrices. Along with the coupling agents referred to above, other methods can be used to enhance the bonding between the hydrophilic natural-fiber surface and hydrophobic polymers. High energy surface treatments (corona discharge or plasma) can create functional bonding groups or other modifications on the fiber surface. Or, for example, alkalization using sodium hydroxide can be used to partially dissolve non-cellulose fiber materials, roughening the fiber surface and encouraging bonding with the matrix; this allows more load-transfer to the fibers, and thus higher mechanical properties of the composite.

3.6.4   Nanocomposites

Various clays are used in polymers to provide specific properties. "Nanoclay" fillers are now used to create nanocomposites, or polymers with small loadings (<5%) of filler particles with nanometerscale dimensions. Organically modified nanoclay fillers can be separated (or "exfoliated") into sheet-like particles with molecular-scale thickness. Dispersed in a polymer, the resulting composite provides unique mechanical and electrical properties for specialized applications, using a relatively simple thermoplastic polymer as the matrix material. Nanoclay fillers are also being developed for use in biopolymers, such as PBAT/PLA blends [67]. In thermoforming applications, nano-filler composites derived from natural minerals with attractive inherent characteristics (e.g., non-toxic, gas barrier properties) have been trialed. Recent testing shows that dispersion of the nanocomposite powder in PP results in the creation of microlayers after thermal stretching. The mineral has polymer-like processing properties which allows for layering in the polymer matrix. Unlike talc or calcium carbonate, the mineral disperses and blends into the polymer. Early data from masterbatch trials show a large reduction in OTR versus PPMA [84].

Technology also exists for isolating cellulose fibers and turning them into nanofillers. Here, a natural fiber's crystalline cellulose component is extracted as microcrystalline particles, or even smaller "nanowhisker" particles, 100–1000 nm long and 560 nm in diameter [85]. These micro- and nano-fillers, when fully dispersed in PLA, can theoretically create bio-nanocomposites of unique properties, but efforts continue towards developing the proper dispersal techniques and coupling agents.

3.7   Concluding Summary

Many key points that relate to plastics and sustainability issues can be synthesized from even this broad overview of polymers and plastics – a concise list of five points is offered below:

1. Common commodity polymers that are simple in structure tend to have smaller environmental footprints, requiring

synthesis processes with fewer steps and less complex constituents, and requiring less production energy.

2. Because of the reactivity of their component chemistries, the "eco-acceptability" of PVC and more complex engineering polymers such as polycarbonate are harder to argue for in terms of sustainability.

3. Large-scale plastics recycling may continue to be limited to simpler, high-volume polymers. More complex polymers, multi-component plastic compounds, or thermosetting polymers and composites complicate most recycling processes, and many are essentially impossible to recycle economically themselves.

4. Biologically synthesized plastics are uniquely suited for food-use applications where their biodegradability can facilitate sending food waste to compost. The reduced carbon footprint is increasingly important when considering climate change impacts, but there is still work to be done growing economies of scale and access to industrial composting infrastructure.

5. Plant-based fiber reinforcement and bio-based additives are useful options (with some limitations) for increasing the "green" content of reinforced plastic compounds. Despite technical and economic challenges, much work continues in this area driven by environmental concerns.

References

1. Anastas, P.T., Bickart, P.H., Kirchhoff, M.M., *Designing Safer Polymers*, Wiley-Interscience, New York, 2000.

2. NatureWorks LLC, Technical resources, http://www.nature worksllc.com/ Technical-Resources, 2011.

3. Yield10 Biosciences (Mirel), Data sheets, 2011, revised 2015. https://www. yield10bio.com/sites/default/files/P1003_Product_Data_sheet-rev_151201. pdf

4. Narayan, R., Bioplastics: Next generation polymer materials for reducing carbon footprint, and improving environmental performance, [Webinar], in: *Society of Plastics Engineers and Ramani Narayan*, 2008, June 4.

5. Huang, H., Microbrial ethanol, its polymer polyethylene, and applications, in: *Plastics from Bacteria: Natural Functions and Applications*, G.-Q. Chen (Ed.), Springer-Verlag, Berlin Heidelberg, 2010.

6. Jem, K.J., van der Pol, J.F., de Vos, S., Microbial lactic acid, its polymer poly(lactic acid), and their industrial applications, in: *Plastics from Bacteria: Natural Functions and Applications*, G.-Q. Chen (Ed.), Springer-Verlag, Berlin Heidelberg, 2010.

7. Pilz, H., Brandt, B., Fehringer, R., *The impact of plastics on life cycle energy consumption and greenhouse gas emissions in Europe (Report)*, denkstatt GmbH, Vienna, 2010, June.

8. Esposito, F., Brazil's Braskem adding PE, PVC capacity, in: *Plastics News*, p. 9, 2011, April 18.

9. Moore, S., Braskem launches renewable EVA derived from sugarcane, PlasticsToday.com, 2018, August 1.

10. 25th Annual National Post-Consumer Plastics Bottle Recycling Report, American Chemistry Council & Association of Plastics Recyclers, 2015.

11. Toto, D., Envision Success, RecyclingToday.com, 2013, December 4.

12. Paben, J., Lack of demand could sink ocean plastics recycling program, ResourceRecycling.com, 2019, January 15.

13. Flexible Packaging Sortation at Materials Recovery Facilities. *Am. Chem. Counc.*, 2016, September 22.

14. Report from the Commission to the European Parliament and The Council on the impact of the use of oxo-degradable plastic, including oxo-degradable plastic carrier bags, on the environment, 2018, January 16.

15. Quoted from www. european-bioplastics.org/eu-takes-action-against-oxo-degradable-plastics/

16. Andreesen, C. and Steinbuchel, A., Recent developments in non-biodegradable biopolymers, in: *Precursors, production processes, and future perspectives*, SpringerLink, 2019.

17. Johnson, J., Closed Loop Fund backing PureCycle project, www.plastics-news.com, 2018, June 11.

18. Company announcements, websites, www.milliken.com, www.purecycle-tech.com

19. Griswold, M., New plastics scorecard favors bio-based resins, in: *Plastics News*, http://www.plasticsnews.com, 2009, October 1.

20. Tabone, M.D., Cregg, J.J., Beckman, E.J., Landis, A.E., Sustainability metrics: Life cycle assessment and green design in polymers. *Environ. Sci. Technol.*, 44, 8264–8269, 2010.

21. Polyvinyl chloride production in the United States from 1990 to 2017, www.statista.com

22. ICIS news, Outlook 2018: US PVC bullish in new year on rising demand, 2018, January 1.

23. https://www.federalregister.gov/documents/2012/04/17/2012-6421/national-emission-standards-for-hazardous-air-pollutants-for-polyvinyl-chloride-and-copolymers

24. ECPI responds to misleading statements about plasticizers, in: *Compounding World*, vol. 6, 2011, April.

25. Ouchi, Y., Yanagisawa, H., Fujimaki, S., Evaluating Phtalate Contaminant Migration Using Thermal Desorption Gas Chromatography-Mass Spectrometry (TD-GC-MS). www.mdpi.org. *Polymers*, 11, 4, 683, 2019.

26. Reinecke, H., Navarro, R., Pérez, M., Gómez, M., Safer polyvinylchloride with zero phthalate migration, (10.1002/spepro.003028). *Soc. Plast. Eng. Plast. Res. Online*, http:// www.4spepro.org, 2010.

27. Polystyrene production in the United States from 1990 to 2017. www.statista.com

28. Ruckdäschel, H., Hingmann, R., Ruch, J., Hahn, K., Development of polymer foams – trends & sustainability, ANTEC 2011 [Proceedings]. *Soc. Plast. Eng.*, 2011.

29. Accelerating Circular Supply Chains for Plastics, in: *Closed Loop Partners*, 2018.

30. https://resource-recycling.com/plastics/2017/05/17/nyc-moves-reinstate-foam-ban-new-analysis/

31. Liu, H., Ou, X., Zhou, S., Liu, D., Microbial 1, 3-propanediol, its copolymerization with terephthalate, and applications, in: *Plastics from Bacteria: Natural Functions and Applications*, G.-Q. Chen (Ed.), Springer-Verlag, Berlin Heidelberg, 2010.

32. US 2551731A, Polyesters from heterocyclic components, 1951.

33. Company filings, presentations, www.avantium.com

34. Report on Postconsumer PET Container Recycling Activity, in: *Association of Plastics Recyclers*, 2017.

35. European PET Bottle Platform (EPBP). www.epbp.org

36. Damm, P.O. and Matthies, P., Nylon, in: *Handbook of Plastic Materials and Technology*, I.I. Rubin (Ed.), John Wiley & Sons, New York, 1990.

37. Mapleston, P., New technologies for a greener industry. *Plast. Eng.*, 64, 1, 10–14, 2008, January.

38. Rhodia Group, Rhodia enriches its product offering with an economical new ecological and organically sourced polyamide [Press Release], http://www.rhodia.com, 2010, 2009, November 26.

39. https://www.energy.gov/eere/amo/nylon-carpet-recycling

40. Pillichody, C.T. and Kelley, P.D., Acrylonitrile-butadiene-styrene (ABS), in: *Handbook of Plastic Materials and Technology*, I.I. Rubin (Ed.), John Wiley & Sons, New York, 1990.

41. Hengstler, J.G., Foth, H., Gebel, T., Kramer, P.J., Lilienblum, W., Schweinfurth, H., Völkel, W., Wollin, K.M., Gundert-Remy, U., Critical evaluation of key evidence on the human health hazards of exposure to bisphenol A. *Crit. Rev. Toxicol.*, 41, 4, 263–91, 2011, April.

42. Tolinski, M., Thermosets stay forever young. *Plast. Eng.*, 64, 2, 6–9, 2008, February.

43. Styrene allies to recommend lower PEL, vol. 27, pp. 14–15, Composites Manufacturing, 2011, Jan.-Feb.

44. Stewart, R., *Styrene watch*, vol. 29, pp. 25–26, Composites Manufacturing, 2011, May.

45. https://www.compositesworld.com/news/aditya-birla-chemicals-acquires-recyclable-thermoset-technology-from-connora-technologies
46. Fraunhofer shows easier-to-recycle thermoset and European Plastics News, http://www.plasticsnews.com, 2011, May 31.
47. Bioplastics Market Data 2018; Global Production Capacities of Bioplastics 2018-2023. European Bioplastics Report.
48. Vink, E.T.H. and Davies, S., Life Cycle Inventory and Impact Assessment Data for 2014 Ingeo Polylactide Production. *Ind. Biotechnol.*, 11, 3, 167–180, 2015.
49. Vink, E.T.H. and Davies, S., Life Cycle Inventory and Impact Assessment Data for 2014 Ingeo Polylactide Production. *Ind. Biotechnol.*, 11, 3, 167–180, 2015.
50. https://packagingeurope.com/low-carbon-footprint-of-pla-confirmed-by-peer-reviewed-life-/
51. Broeren, M., Kuling, L., Worrel, E., Shen, L., Environmental Impact Assessment of Six Starch Plastics Focusing on Wastewater-Derived Starch and Additives. *Resour. Conserv. Recycl.*, 127, 2017 December 2017. Elsevier.
52. Yu, L. and Chen, L., Polymeric materials from renewable resources, in: *Biodegradable Polymer Blends and Composites from Renewable Resources*, L. Yu (Ed.), John Wiley & Sons, Hoboken, NJ, 2009.
53. Wong, S. and Shanks, R., Biocomposites of natural fibers and poly (3-hydroxybutyrate) and copolymers: Improved mechanical properties through compatibilization at the interface, in: *Biodegradable Polymer Blends and Composites from Renewable Resources*, L. Yu (Ed.), John Wiley & Sons, Hoboken, NJ, 2009.
54. Kochesfahani, S.H., Abler, C.A., Crépin-Leblond, J., Jouffret, F., Enhancing biopolymers with high performance talc products, ANTEC 2010 [Proceedings]. *Soc. Plast. Eng.*, 2010.
55. Carlin, C., Innovation Takes Root. *Conference Report, Plastics Engineering*, vol. 74, Issue 10, 2019.
56. Chen, G.-Q., Plastics completely synthesized by bacteria: Poly-hydroxyalkanoates, in: *Plastics from Bacteria: Natural Functions and Applications*, G.-Q. Chen (Ed.), Springer-Verlag, Berlin Heidelberg, 2010.
57. Schut, J., Where is Metabolix's PHA biopolymer? *Plast. Eng. Blog*, 2011, 2011, January 19. http://plasticsengineeringblog.com
58. https://danimerscientific.com/2019/11/12/danimer-scientific-and-genpak-partner-to-launch-new-line-of-biodegradable-food-packaging/
59. Krishnaswamy, R.K., Kridaratikorn, S., Kann, Y., Bhoyar, R., McCarthy, S., Kalika, D.S., Smith, P., Characterization of the microstructure of poly (hydroxy butanoic acid) copolymers during their post-fabrication annealing at room temperature, ANTEC 2010 [Proceedings]. *Soc. Plast. Eng.*, 2010.
60. Jacquel, N., Lo, C.-W., Wei, Y.-H., Wu, H.-S., Wang, S.S., Isolation and purification of bacterial poly(3-hydroxyalkanoates). *Biochem. Eng. J.*, 39, 15–27, 2008.

61. Jung, Y.K., Lee, S.Y., Tam, T.T., Towards systems metabolic engineering of PHA producers, in: *Plastics from Bacteria: Natural Functions and Applications*, G.-Q. Chen (Ed.), Springer-Verlag, Berlin Heidelberg, 2010.

62. Thellen, C. and Ratto, J.A., The use of biobased/biodegradable materials in food packaging, ANTEC 2011 [Proceedings]. *Soc. Plast. Eng.*, 2011.

63. Lawton, D., Wang, G., Liu, Q., Thompson, M.R., Twin screw extrusion of thermoplastic potato starch, ANTEC 2010 [Proceedings]. *Soc. Plast. Eng.*, 2010.

64. Yuan, H., Liu, Q., Hrymak, A., Thompson, M., Ren, J., Thermoplastic potato starch blends and bioplastic films. ANTEC 2010 [Proceedings]. *Soc. Plast. Eng.*, 2010.

65. Grewell, D., Carolan, S.T., Srinivasan, G., Enhanced water stability of soy protein plastics using acid anhydrides, ANTEC 2010 [Proceedings]. *Soc. Plast. Eng.*, 2010.

66. https://footwearnews.com/2019/business/manufacturing/algix-bloom-foam-sustainability-algae-materials-1202786956/

67. Shahlari, M. and Lee, S., Polymer-clay nanocomposites of poly(butylene adipate-co-terphthalate) and poly(lactic acid) blend: 1. Effects of incorporating organically modified silicate layers and the mixing sequence on the morphological and rheological properties, ANTEC 2010 [Proceedings]. *Soc. Plast. Eng.*, 2010.

68. Li, G. and Favis, B.D., Morphology development and interfacial interactions in polycaprolactone/thermoplastic starch blends. ANTEC 2010 [Proceedings]. *Soc. Plast. Eng.*, 2010.

69. Industry news and notes. *Plast. Eng.*, 66, 8, 54, 2010, September.

70. Tolinski, M., Additives annual. *Plast. Eng.*, 2010, November/December.

71. Tolinski, M., *Additives for Polyolefins*, Elsevier/Plastics Design Library, Oxford, 2009.

72. https://polymer-additives.specialchem.com/selection-guide/plasticizers, 2019

73. https://www.compositesworld.com/blog/post/the-state-of-recycled-carbon-fiber, 2019

74. https://www.compositerecycling.org, 2019

75. Biron, M., Plastic reinforcement – What's stopping us from going the natural way? http://www.specialchem4polymers.com, 2010, May 17.

76. Tolinski, M., Glass fiber meets stiffer competition. *Plast. Eng.*, 61, 3, 18–20, 2005, March.

77. Thitithanasarn, S., Leong, Y.W., Hamada, H., Effect of natural fiber treatment on the interfacial adhesion and mechanical performance of poly (lactic acid)-based textile insert moldings, ANTEC 2010 [Proceedings]. *Soc. Plast. Eng.*, 2010.

78. Wakabayashi, K., Vancoillie, S.H.E., Assfaw, M.G., Desplentere, F., Van Vuure, A.W., Flax Fiber-Polyamide 6 Composites via SSSP: Expanding the Portfolio

of Natural Fiber Reinforced Thermoplastics", ANTEC 2017 Proceedings. *Soc. Plast. Eng.*, 2017.

79. Hardy, P.A., Lee, E.C., Simon, L.C., Commercialization of injection moldable composites utilizing wheat straw fiber, Global TPO Conference [Presentation]. *Soc. Plast. Eng.*, 2010, October 6.

80. Shah, A.U., Sultan, M.T.H., Jawaid, M., Cardona, F., Abu Talib, A.R., A Review on the Tensile Properties of Bamboo Fiber Reinforced Polymer Composites. *BioResources*, 11, 4, 10654–10676, NC State University, 2016.

81. http://www.fao.org/economic/futurefibres/fibres/abaca0/en/, 2019.

82. Moriana, R., Karlsson, S., Ribes-Greus, A., Reinforced bio-composites with guaranteed degradability in soil (10.1002/ spepro.003064). *Soc. Plast. Eng. Plast. Res. Online*, http://www.4spepro.org, 2010.

83. Greer, D.S., Bradley, W.L., Natividad, D., Vano, R.J., More sustainable non-woven fabric composites for automotive using coir (coconut) fibers, Automotive Composites Conference & Exposition [Proceedings]. *Soc. Plast. Eng.*, 2010.

84. Ouchchen, A., Ph.D, High Barrier Nanolayer Composition to Replace Multilayer Packaging. *Conference Proceedings, AMI Thin Wall Packaging*, Chicago, 2017.

85. Stoeffler, K., Ton-That, M.T., Denault, J., Luong, J., Wu, C., Sain, M., Polylactide composites filled with microcrystalline cellulose, nanocrystalline cellulose and cellulose nanofibers, ANTEC 2010 [Proceedings]. *Soc. Plast. Eng.*, 2010.

4

Applications: Demonstrations of Plastics Sustainability

Major plastics consuming industrial sectors have all tried various uses of bio-based and/or recycled plastics. Some of these applications help create benchmarks against which one can predict what might be possible in the future. While others are dead ends that offer few environmental benefits over the use of conventional plastics. In most cases, it is still too early to tell which kinds of bio-based/recycled applications will be economically feasible over the long term. Factors such as fossil fuel prices, environmental regulations, commercial feasibility, or simply consumer tastes will determine the future, but are hard to predict. Yet the past decade has shown dynamism, with some companies failing and new ones appearing.

This chapter will strive mainly to discover the factors that determine which environmentally sustainable plastics applications are themselves truly sustainable and important, and which are not. The following plastics consuming market sectors and trends will be overviewed:

Michael Tolinski and Conor P. Carlin. Plastics and Sustainability 2nd Edition: Grey is the New Green: Exploring the Nuances and Complexities of Modern Plastics, (127–158) © 2021 Scrivener Publishing LLC

- Sustainable plastics application trends in general (4.1).
- Packaging applications for bio-based and recycled plastics (4.2).
  - Traditional plastic bags and containers: use, disposal, and recycling
  - Bio-based plastics for packaging
  - Greener foams for packaging
- Sustainable plastics in building and construction (4.3).
  - Recycled/recyclable construction applications
  - Wood-plastic composites
- Automotive plastics and sustainability (4.4).
  - Lightweight plastics for improving fuel efficiency
  - Recycled plastics and end-of-vehicle-life issues
  - Bio-based plastics in the automotive industry
- Specialized applications and plastics sustainability (4.5).
  - Electrical/electronics applications
  - Medical plastics and packaging
  - Agricultural applications
- Conclusions about sustainable plastics applications (4.6).

The Coffee Capsule: From Hero to Zero

Among all disposable plastic items, the single-serve coffee capsule is arguably most emblematic of today's convenience-driven economy. It is also notorious for being almost entirely unrecycled, despite claims and proof of recyclability. Originally invented by college roommates in Massachusetts, Peter Dragone and John Sylvan, in 1990, Keurig (from the Dutch word for 'excellence') was developed to allow people to have a single cup of coffee instead of making entire pots. The first machine-pod combination was launched in 1998. The growth of Keurig coincided with an overall increase in caffeine culture and perhaps coffee snobbery, e.g. $7 for a cup in boutiques such as Blue Bottle. In Europe, the preference for espresso led to similar developments in single serve technology with companies such as Nespresso and Lavazza leading the way.

The year 2012 marked an important milestone for coffee capsules: the patents on Nestle and Green Mountain/Keurig products expired. After years of continued growth, the market has moved from a duopoly to something more fragmented, where a sub-industry of compatible capsules has emerged. Still, with approximately 50 billion capsules produced annually, there is double-digit growth in all geographic regions, including 20-25% in NAFTA and close to 100% in Latin America. The current material breakdown is ~70% plastic and ~30% aluminum, with some interesting developments in cellulosic-based materials [1].

Advances in material science have played a critical role in coffee pods which must protect the taste and aroma of coffee grounds while also being able to withstand high temperatures and pressures. Economies of scale have developed to allow pods to be purchased for as low as $0.35 each. The European tradition of short, intense, espresso beverages places distinct requirements on the package which contains 6g/0.22oz of coffee and has to withstand a brewing temperature of around 90°C and brewing pressure of 19bars. By comparison, the typical US-based K-Cup contains about double the coffee (12g/0.45oz) with much less pressure. Both markets see plastic-based capsules with low oxygen transmission rates (OTR) for shelf life and aroma preservation. The vast majority of NAFTA-based capsules are thermoformed with high-barrier polymer technologies, primarily styrene-based materials with EVOH layers. There has been a well-documented move toward PP in recent years, with Green Mountain pioneering a new co-injected PP version of its K-Cup. Other technologies include monolayer capsules with barrier IML or coated with silicone oxide (SiOx).

With consumer pressure growing, however, there has also been a notable increase in bio-based material developments that seek to address the end-of-life concerns. NatureWorks in the US and Flo in Italy are two examples of companies striving to develop a compostable pod that can withstand the brewing rigors and disintegrate in the correct environment. PLA and cPLA pods have recently been introduced on supermarket shelves. Other start-up companies on both sides of the Atlantic continue R&D efforts in non-plastic materials including paper/pulp and aluminum. As with most any consumer good, however, human behavior plays an outsized role in what actually happens to any item when its useful life has expired. Without access to industrial composting facilities, are compostable pods really any better for the environment? Is home composting a viable option? What about take-back programs offered by Nestle for their used aluminum Nespresso pods? These and other thorny questions remain as we continue to savor the flavor of our caffeine habit.

4.1   Trends in Sustainable Plastics Applications

From 2010–2020, bio-based and recycled-content plastics achieved a new level of commercial-scale production that has allowed their wider use in various common applications, most of which have traditionally used fossil fuel-based virgin resins. Now, most grocery stores carry transparent salad trays or egg cartons labeled as being made from PLA or as having post-consumer PET content. Some snack-food bags are also being made from PLA, replacing polyolefins. The polyurethane foam in some car seating is now partially based on soy-based resin. And more decking, fencing, or park benches are being made from post-consumer plastics used in wood-plastic composites.

During these years, developers of sustainable plastics applications have been recognized with environmental awards from numerous plastic and packaging industry associations, as well as consumer goods groups. Automotive components, laptops, luxury brand packaging, and carpeting tile are just a few areas where new products are the results of scores of small, iterative breakthroughs in many areas of plastics processing technologies.

These types of applications and material developments are over-coming some obstacles that have limited the sustainable uses of plastics. Plastics are indeed often difficult to separate from used products for recycling. They are then hard to reuse in products with challenging property requirements. And many engineering plastics with complex chemistries usually cannot be bio-based or made from biopolymers. The above developments have attacked these challenges, but the developments also give rise to general, abstract questions, such as:

- How much does a *partially* bio-based plastic count as a sustainable product?
- How much food crop production should be devoted to making bio-based plastics?
- When does it make economic sense to recover post-consumer plastics from certain products?
- How much can we ever truly "close the loop" for plastics, and use post-consumer plastics in their original or more rigorous applications?
- And when do the valuable weight- or resource-savings qualities of plastic products compensate enough for the impacts of the resources used in their production, or difficulties with their recycling or disposal?

The applications in key plastic application sectors covered below – packaging, construction, and automotive, plus a few key specialized applications – at least indicate recent and future progress towards finding answers to these questions.

4.2   Sustainable Plastics Packaging

Over the past few decades, plastics have won a greater share of packaging applications as primary packaging for all kinds of products, especially food and beverages. Packaging is the largest end-use sector for packaging, out-growing substitute materials such as glass and cartons. The global volume share for the top 20 types of packaging in 2018 shows plastics make up over 50% of all packs made [2]. The percentage will likely get higher; after all, glass packaging is easily damaged and heavy; paper fiber materials lack transparency and water resistance, and metal packaging has aesthetics, damage, and weight issues as well. Moreover, all three of these categories of materials are energy-intensive to produce. In contrast, commodity plastics packaging is lightweight, impact-resistant, and water- or gas-impermeable in the various degrees required – plus it can be transparent, tinted, or opaquely colored, with bright, glossy surface aesthetics.

These qualities have contributed to plastics' steep rise in packaging. But the plastics industry has also faced an equal amount of increasing complaints about the waste stream created by single-use plastics packaging. Critics point out that products are often over-packaged, and require consumers to dispose of this excess material if the plastic cannot be accepted for recycling. This direct handling required by the consumer to dispose of plastics packaging is perhaps why many consumers claim to dislike plastics in general – in the case of packaging, plastics are simply seen as unnecessary waste. These consumers do not likely regard these disliked plastics as being in the same category that includes other, durable plastics applications that they depend on in their clothing, automobiles, electronic devices, carpeting, and so forth.

Thus, "plastics" to some critical consumers, simply means "plastic packaging I must regularly throw out." For example, consider the case of a magazine publisher encouraging its staff to pursue a "Plastics-Free February" during which staff members were to avoid purchasing or using plastic products. Yet one of the publisher's own consumer publications is a sports magazine about running – a publication that depends on running articles and advertising that promotes all kinds of accessories that serious runners rely on. Nearly all of these accessories use various polymer fibers and polymeric materials in their construction, ironically showing how "bad" and "good" plastics are categorized in people's minds [3].

Given consumer concerns (or confusion), plus the genuine concern that too much plastics packaging does indeed become waste, packaging plastics will be the crux or focal point for efforts in shifting to greener plastics: recycled, recyclable, biodegradable, bio-based, or some combination of these. Meanwhile, given both plastics' prevalence and short-term uses in so many packaging applications, more calls for sustainable packaging can be expected in the future.

Because so many consumers are concerned about the environment, they will become more interested in knowing if or what part of a packaged product is eco-friendly. But currently, bio-based and sustainable packaging comes with a cost premium due to the novelty and relatively small economies of scale of bio-based and recycled materials. Ultimately, it may be the consumers who drive major developments in sustainable packaging, since packaging manufacturers tend to stay focused on minimizing costs and staying competitive in challenging economic times. Consumer behavior is, of course, also driven by price, but by agreeing to pay a certain premium for a sustainably packaged product, at least a small percentage of consumers may help determine what kinds of packaged products will start to incorporate bio-based or recycled materials.

Large corporate users of packaging or sellers of common packaged products have noticed the cultural and consumer trends in green thinking, and are responding. Large consumer products producers such as Nestle, Unilever, and Procter & Gamble have announced a long-term sustainability initiatives, both individually and in partnership with the Ellen MacArthur Foundation's "New Plastics Economy Global Commitment". The priority is to use only packaging made from recyclable or renewably produced materials, with a goal of eventually eliminating all fossil-fuel-based virgin plastic use. As one of the major global consumer goods companies (and also the world's largest B-corporation), Danone is a signatory member of the New Plastics Economy Global Commitment. The company is taking a 'glocal' approach to packaging, engaging governments and regulatory bodies at the global level to shape policy, and working with regional partners to act locally. Similar to UK-based retailer Tesco and other large entities, Danone aims to make 100% of its packaging reusable, recyclable, or compostable by 2025. Currently, the company produces almost 1.6 million tons of packaging per year, of which approximately 50% is plastic and 33% is paper. One-third of this is made with recycled material and 50% is recycled. As a large producer of dairy goods, especially yogurt, Danone faces an imminent materials challenge with PS, though the company has not yet (at least publicly) decided what it will do. From 1978 until 2012, there was no significant change in PS for dairy producers. Conversely, PET – as exemplified by the ubiquitous water bottle – continues to evolve going from virgin PET to mechanical rPET, to bio PET, to molecularly recycled rPET [4].

But the term "sustainability" will continue to be read by consumers mostly without understanding certain distinctions. Sustainable packaging may be plant-based, recycled, recyclable, biodegradable, reduced in material content, or some combination of descriptors. The multiple interpretations of the term "sustainable" could be an advantage, creating different routes to drive new consumer-driven green packaging. Yet the industry also has legitimate concerns about consumers favoring sustainable packaging that is labeled as such but is not truly sustainable or even recyclable. Hence, the continuing calls for better standards on what products can truly be considered sustainable packaging, including the development of the "How2Recycle" labels created and managed by the Sustainable Packaging Coalition (see Figure 4.1 for example).

In their desire to reduce costs, plastics packaging manufacturers already are pursuing one key aspect of sustainability: using less material, while still fulfilling the same packaging requirements. Designing packaging with thinner dimensions (a topic covered more in Chapter 5) comes naturally to the industry. For example, PET water bottles have become increasingly thinner and lighter over the years, yet unintended consequences of lightweight pose technical and economic challenges for recyclers. Computer-controlled process equipment has allowed these major reductions in

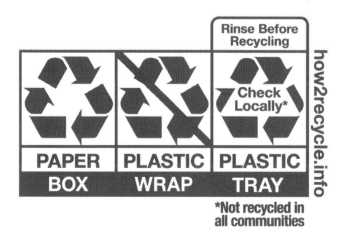

Figure 4.1 Sample "How2Recycle" label developed by the Sustainable Packaging Coalition.

material use. However, designers will eventually encounter a natural limit to how thin packaging material can be for an application before its products are no longer adequately protected, or the consumer interprets the packaging as being too flimsy.

In the next sections, the packaging focus will be narrowed to plastic shopping bags and rigid food and beverage containers: the two basic high-volume applications in packaging where issues about fossil-fuel-based plastics, recycling/disposal, and bio-based/biodegradable plastics converge.

4.2.1   Plastic Bags and Containers

According to the mass media at least, consumers hate plastic bags (mostly PE) and bottles (mostly PET), though their behavior shows that they love using them. Possibly because of this contradiction, these packaging products have taken on a strange role – they are almost like scapegoats that symbolize all the evils of an over-consuming society that feels somewhat guilty about its tendencies. Other than the key fact that bags and bottles tend to be improperly disposed, there is little rational argument in the efforts to discourage or prevent their use. There is not nearly as much discussion about limiting the numbers of miles people drive per year (a much higher energy-consuming activity that has the potential to change the earth's climate relatively soon), or about how much paper packaging people use (a material which requires more energy to produce and transport than plastic), or about how often people purchase new cell phones or electronic devices (and how they dispose of their old ones). Yet plastics producers will always have to deal with political trends and consumer choices that are beyond rational explanation.

That being said, it is still important to look at the rational side of the argument: the improper disposal of single-use plastics. Reliable, current sources are difficult to identify for total amounts, though one report states approximately 380bn plastic bags and wraps are produced each year in the US alone [5]. Globally, one can estimate multiple factors of this figure when considering the populations of China, India, and ASEAN countries. And common sense about human behavior dictates that at least several tons of these bags ended up as litter. This litter is persistent in the environment, and the polyolefin component of it is less dense than water, meaning it can float down streams to rivers, and eventually to the ocean where it creates major problems for aquatic life. Obviously littering is the kind of human behavior that should be focused on for modification. But even with greater enforcement and education against littering, a noticeable proportion of bags and bottles will still end up on the sides of roads and in waterways.

One way offered to address the waste issue is the use of biodegradable plastic packaging. Yet questions remain about how completely these kinds of biodegradable compounds degrade in various environments after becoming litter. By contrast, inherently biodegradable biopolymers (discussed more below) show complete degradation, at least in composting conditions, over a preferred time scale of a few months rather than years.

For now, other than placing into effect even more outright bans on some plastic packaging, the best solution is to discourage littering and encourage more collection and recycling of the packaging. The recycling rate for bags and film wrap has plenty of room to grow, and has indeed increased by 54% from 2005 to 2017 [6]. Plastic bottle recycling is better, but rates still remain below 30% in the United States and around 50% in Europe. These rates seem especially low when considering the energy value of just PET bottles, yet economics remains the primary driving factor. Continued capacity expansion in all regions of the globe have driven prices downward, creating significant problems for recyclers who face different cost structures than their virgin counterparts [7]. A breakdown of plastic packaging by polymer type is shown in Figure 4.2, suggesting perhaps that efforts for recovery should start with PE.

There is growing interest in using recycled material in more new packaging products, spurred by consumer sentiment and industry initiatives, such as the "Demand Champions" program sponsored by the US-based Association of Plastics Recyclers and the CEFLEX project in Europe. The former seeks

Figure 4.2 Global Plastics Packaging Market (Source: Plastics News).

to address the demand-side of the market equation through voluntary commitments from plastics converters to use more recycled material, while the latter is a collaborative consortium set-up to ensure that flexible packaging recycling becomes viable. Meanwhile, makers of wood-plastic composites use millions of old PE bags and film in decking and fencing products.

But the still-developing markets for recycled bag and bottle materials encounter inherent obstacles. Only so much recyclate can be used in new bags and bottles without excessively weakening their properties. Moreover, if a purified recycled material is to be used in new food and beverage packaging, testing must confirm that it is food-contact compliant, and that no odors or flavor-affecting residuals are released from it. Fortunately, signs of hope come from the small number of PET recyclers that produce rPET with "letter of no objection" status from the US FDA and similar government regulatory bodies, a designation which allows the rPET to be used in new beverage/food containers. New work is underway to develop food-grade recycled PP [8].

A much more common alternative end-use for recycled packaging material are the downcycled applications of fiber or strapping (for PET), or construction applications (for polyolefins), which together consume most recycled packaging resins. Apart from carpet fiber, more rPET is being used for textile fiber as better reprocessing technology creates improvements in material quality.

A final factor that complicates recycling trends is the inherent energy content in recovered plastics itself, meaning that waste-plastics incineration with energy recovery is favored over recycling in some regions. In Europe, for example, recycling overtook landfill in 2016 as an end-of-life option, yet energy recovery was still the main outcome for 41% of all collected plastics [9]. This essentially treats waste plastics like a fossil fuel – though it is a fuel which can require a great deal of energy for its "refining" into a synthesized, compounded, and formed plastic product.

4.2.2   Bio-Based Plastic Packaging

Bio-based plastic packaging is intended to remove the fossil fuel stigma from plastic packaging, attracting some consumers and making others less critical of plastics packaging, while also decreasing the eco-footprint of packaging overall. But much work needs to be done in reducing bioplastics' costs and in confirming that their total life-cycle impacts are indeed less than those of traditional packaging plastics. Especially for bioplastics based on food crops such as corn, the impacts of pesticide and fertilizer use and feed-stock collection and conversion energies must be low enough so that bioresin producers can make a strong environmental case in their favor.

Polylactic acid (PLA) has become the key bioresin of interest for packaging applications. It is compostable, and its properties resemble those of PP, PET, and polystyrene, typically meaning that consumers are not even aware when it is being used in a package unless a label tells them.

Expanded uses of PLA products are still being cautiously weighed against their effects in the current recycling infrastructure. In recycling, PLA's similarities with PET become a disadvantage. PLA acts as a

The Growth of the PLA Ecosystem

"Innovation Takes Root" is a biannual conference hosted by NatureWorks, developer of the Ingeo PLA polymer and the Vercet lactide chemistry platforms. The most recent event, in 2018, attracted over 250 people from 27 countries. A blend of workshops, fireside chats, seminars and keynotes ensured an engaged audience over the three-day period in a location that wasn't limited to windowless, air-conditioned meeting rooms. Major topics included the continued evolution of the Ingeo platform, circular economy principles, regulatory and legislative developments related to bio-based plastics, PLA use in additive manufacturing, and new developments in single-serve capsules.

One topic that engaged a majority of the value chain was new product development for the food service industry. According to figures from the Green Restaurant Association cited by Nicole Whitemann, applications development engineer at NatureWorks, 39 billion pieces of disposable cutlery are produced each year, only 1% of which is recycled. Coated paper cups are included in these figures and though they are generally easy to recycle with other fiber sources, the use of PE or PLA can impact the level of biodegradable content which affects the decision to recycle or compost the items. There is a growing portfolio of examples and case studies illustrating just how much savings can be found via effective waste diversion programs. Marquee programs from the London Olympics to major sporting stadia have shown significant strides toward making zero waste a reality. Whitemann announced the launch of Ingeo 1102, a new PLA extrusion coating grade that exhibits improved performance through optimization of polymer design. Compared to PE, PLA has lower melt strength and higher viscosity which creates manufacturing challenges such as high neck-in and draw resonance that limits line speeds. Ingeo 1102 is designed to balance some of the conflicting attributes of PLA to deliver a more stable web with good adhesion and good seal strength while improving flow and penetration into paperboard to reduce coating weights and improve cup sealing range [10].

Beyond new Ingeo-based product introductions from NatureWorks, several leading players in the biopolymer ecosystem announced developments in additive masterbatches such as impact modifiers for transparency, slip and antiblock agents for coefficient of friction changes, and barrier films for enhanced shelf life applications. Sukano (Duncan, SC) presented data on a variety of additives including their transparent modifier S633 as used in thermoformed packaging. This nucleating masterbatch forms many small nucleation sites which increases the speed of crystallization. A melt strength enhancer improves IV and mechanical properties while retaining transparency. Using LDR of 1-3% is shown to reduce brittleness without affecting temperature resistance. For injection molding grades of PLA, Sukano S687-D, an opaque impact modifier, improves elasticity and toughness for high-stress applications when used at 10-30% dosage rates [11].

contaminant in PET recycling, and requires advanced separation technologies to remove it from the recycled product stream (also see Chapter's 2 and 3 discussions). Meanwhile, the industrial infrastructure for recycling or composting waste PLA has not yet sufficiently developed.

A paradox of sorts exists for PLA producers and users: what do we do with the products at the end-of-life? Though the material meets the requirements found in EN 13432 and can be composted in industrial facilities, it is an acknowledged fact that we do not yet have adequate infrastructure for large scale composting. Compounding this challenge is resin code #7 – "other" – where PLA is lumped with many other polymers including ABS, PC and multilayer films. For a polymer to be given its own resin code, there must be 400MM lbs of production for 3 consecutive years. In addition, 40 state legislatures must be lobbied so that the code can be created and written into law. Because PLA is a still a relative newcomer to the family of thermoplastic polymers, it has not yet reached the levels of production that can trigger the necessary recycling mechanisms. It is clear that the product, development, scale, and ultimate life cycle for new polymers is measured in generations, not years [12].

Another green avenue for packaging is the use of polymers that are identical to traditional polymers except in that they are partially or fully synthesized from renewable feedstocks. This means that although the packaging product is not compostable, it will be perfectly suitable for existing recycling methods.

Materials in the PHA (polyhydroxyalkanoate) family of biologically derived polymers are also being formulated for packaging. PHBV copolymer (poly(3-hydroxybutyrate-*co*-3-hydroxyvalerate) offers properties that make it a feasible though rather high-priced alternative, while also being highly biodegradable. PHBV has also been blended with PLA in efforts to develop better bioplastics for packaging and films with adequate low-temperature (refrigeration) resistance, as well as good stiffness and strength [13].

Meanwhile, bio-based resin packaging faces competition from other types of bio-based materials that are mainly based on raw plant fibers. Molded food service items and tableware are being made from waste sugarcane fibers blended and molded with starch-based resins. Moldable wood pulp fiber is another material being used for sustainable/compostable packaging.

4.2.3 "Greener" Foam Packaging

With improved bioresins, makers and users of polymer foams have great potential to improve their environmental footprint and the public's perception of foamed plastics. Traditional plastic foams are extremely useful for packaging, seating, and insulation, but they also have a poor environmental reputation because they are generally not recycled and take up large volumes as waste materials. However, more bio-based, biodegradable polymer foams are being developed for situations where they are effective.

In recent years, it has mainly been non-packaging foam applications that have adopted bio-based polymer content. Automotive manufacturers such as Ford Motor Company and Johnson Controls Inc. (JCI) and other automotive specialists have been greening-up foams used for automotive seating, improving company reputations by the increased biocontent in vehicles.

However, most foams that are visible to the consumer are packaging foams made from polystyrene, which has a poor environmental image. It is not surprising that expanded PS (EPS) struggles with its environmental reputation, given that its products are rarely recycled and are generally thrown out after a brief use. Durable uses of EPS are important – as in seating or thermal insulation – but these foams are usually hidden behind walls or trim cover materials and are not in the public consciousness.

The recycling of EPS has been taken more seriously lately, even though its low density makes it extremely economically inefficient to transport to recycling facilities. One company in particular, Agilyx, (Portland, Oregon) has commercialized technologies to break PS foam back down to its monomer components. A major supplier of foam products, Dart Container Corporation, has established sites to collect post-consumer PS foam, in response to pressures in communities to ban or recycle EPS food service containers. However, here the "loop" is not totally "closed," since the foam cannot be reused for food-contact items. Dart has faced enormous challenges trying to convince the public and municipalities that its products can be recycled. A high-profile ban on EPS containers in New York City in 2015 led to increased public scrutiny of the material, despite Dart's attempts to illustrate its efforts to recycle, including a network of take-back and drop-off locations. In 2020, the city of San Diego suspended enforcement of its ban following challenges from Dart about environmental impact studies [14].

Given plastic foam's difficulties with recycling, foams made from a bioresin such as PLA have been proposed as a more sustainable choice. However, PLA is generally difficult to foam. To improve PLA's melt strength for foaming, polymer chain-modifying additives can be used to create foams with uniform closed cells. Since PLA is not a totally amorphous polymer like PS, control of PLA crystallization is also critical for creating the desired cell structure. Studies comparing the way PS and PLA foams form are leading to a greater understanding of how to create more useful PLA foams [15].

4.2.4   Key Points

A few key points can be made about present and future trends towards a reduced environmental impact of plastic packaging:

1. The worldwide volume demand for plastics packaging will not decrease anytime soon, but rather the opposite is happening. This, coupled with environmental concerns about plastics' litter and nonrenewable resource use, will maintain or increase the pressure on the industry to develop better, cheaper, bio-based plastic materials for packaging, as well as increase calls for more recycling of packaging.

2. The recycling and recycled content of plastics packaging can be increased, though recycling likely will face natural limits in the percentage of used packaging that can be collected and reclaimed, and in the percentage that can be used in packaging requiring certain properties.

3. Bio-based packaging resins are slowly developing a following of retailers and consumers who are interested in

sustainably-packaged products. These materials' integration in the marketplace is also being driven by consumer products producers and retailers interested in projecting a green image by adopting more bio-based packaging.

4.3   Sustainable Plastics in Building and Construction

As the second largest end-use sector for plastics (consuming around 18% of all plastics used), building and construction applications consume a large variety of plastic types and forms [16]. Plastics are used in construction in nearly the exact opposite way as how they are used in the packaging sector, in that:

- construction plastics are meant for long-term use, rather than relatively brief, disposable uses;
- construction applications are often bulky and space consuming rather than small, lightweight, and portable;
- unlike packaging, many construction plastics uses are not visible to the consumer because they are hidden behind walls or buried underground, as in wire and cable coverings and piping; and
- plastic construction products are invisible in another way – they are often intended to mimic other materials such as wood or stone, drawing attention away from the fact that they are plastic.

This last point perhaps is not surprising, because still in the minds of most people are positive associations about the look and feel of wood in a dwelling or office environment – a warmth, comfort, and natural texture that until only relatively recently plastics have been unable to emulate. Appropriately for construction and building applications, the "bio" content of plastics has mostly to do with the use of real wood fillers that help plastic composites mimic the density and textures of solid wood. The bulky, high-volume nature of many construction applications also serves as a high volume end-use for recycled plastic from packaging applications.

Recent trends are positive for the highest-volume construction applications for plastics. Decking and siding are major plastics consuming products, along with fencing, railing, flooring and carpeting, shingles and roofing, window frames, and wall coverings. Plastics are very visible as siding, where despite its poor image in consumer products applications, vinyl (PVC) remains the plastic of choice over 35 years of use. PVC now makes up at least 50% of the entire siding market [17], up from 33% in 2010, and polypropylene wood-plastic composites (WPC) are also entering the siding market. Polyethylene and other commodity resins are commonly used in WPCs for decking and related composite lumber. For decking and other applications, the WPC share of US market demand was expected to grow at 5% CAGR through to 2025, according to reports from Mordor Intelligence [18]. Driving some of these trends is increasing interest in "green building" – following guidelines which specify the use of more materials with recycled or renewable content, or the use of parts that reduce energy use in buildings. For example, the US EPA's Environmentally Preferable Purchasing Program specifies that buyers purchase construction materials with minimum bio-based or recycled content, such as:

- plastic lumber composite panels with 23% minimum bio-based content;
- plastic shower and restroom dividers and partitions with at least 20% post-consumer recovered material;
- non-pressure PVC pipes containing at least 25% recovered PVC overall, and 5–15% post-consumer content;
- and plastic floor tiles containing 90–100% recovered plastic [19, 20].

In 2000, a methodology for rating green building practices was established by the US Green Building Council (USGBC), called Leadership in Energy and Environmental Design (LEED). The LEED system allows builders' practices to be scored according to six categories, most of which relate at least in some way to plastics' use in construction, including the categories Materials and Resources, Energy and Atmosphere, Indoor Environmental Quality, and Innovation of Design [21].

Plastics are being used to increase LEED ratings in new buildings. Plastics can be formulated to last longer than wood, resisting rot, water, and UV light damage, thus requiring replacement less often. Plastic foams can provide better thermal insulating qualities than other materials. And siding, carpeting, and wood-plastic composites can be made using recycled plastics, with WPCs also using sawdust, a renewable/bio-based waste material.

Fortunes may even be turning for certain previously discouraged uses of plastics in construction. For example, the original LEED rating system in 2000 rewarded builders for avoiding using PVC in new buildings. Since then, the USGBC itself presented a study that analyzed the total environmental and health impacts of PVC in four construction product groups. The study determined that, except in the case of resilient flooring, PVC may still be a better or acceptable choice for siding, window frames, and drain/waste/vent pipes [22].

4.3.1   Recycled/Recyclable Construction Applications

Plastic construction products tend to consume recycled materials rather than serve as sources of it. Along with WPCs that contain post-consumer recycled polyolefins and waste wood, even PVC siding has been commercialized with up to 60% recycled vinyl content, showing a 40% increase in PCR since 2014 [23]. Recycled polyolefins are also being reinforced with glass fiber for durable structural applications, such as lightweight composite bridges, marine structures, or railroad ties (see sidebar about QRS in Chapter 3).

Some work still must be done, however, to find economical uses for construction plastics when they reach the end of their service lives. To this end, old vinyl flooring, pipes, window profiles, and other uses have been researched as potential feed streams for recycled PVC. This PVC can be recycled using mechanical or chemical methods, potentially giving the material a reputation boost by reducing its life-cycle impact. However, the relatively low costs of landfilling old vinyl and the low price of virgin PVC (which make it so attractive for construction in the first place) both deter the creation of an infrastructure for collecting and reprocessing old PVC.

Recently, at least one waste building product has received more positive attention for both its economical use and source of recycled plastic: old carpeting. In the United States, a combined effort by various cooperating companies has demonstrated how old carpeting can be collected and recycled at high volumes. The partnership, called Carpet America Recovery Effort (CARE), was started in 2002 as an industry network of several dozen companies working together to redirect old carpeting away from landfills and into new recycling facilities. Old carpets are normally landfilled because there has never been an economical collection and reclamation infrastructure. Thus over 2 billion kg of carpeting are landfilled in the United States each year, roughly half of which is recoverable, valuable, nylon face fiber [24]. The other half of a carpet's composition is polypropylene backing and calcium carbonate-filled latex that holds the fibers in the carpet – these materials can also be recovered.

Special recycling facilities at companies such as carpet producer Mohawk Group (one of the world's largest bottle recycling companies) have reportedly been able to recover about 30–40% of useful plastics from each kilogram of discarded carpeting. They also use recovered rubber from tires to develop new rubber-backed carpet products such as flooring mats.

Still only about 5% of discarded carpeting is being diverted from landfills. But this proportion is about 400% higher than it was in 2002, and the recovery rate did not decrease even through the Great Recession [25]. CARE's most recent report states that a total of 5.6 billion pounds of carpet has been diverted from landfills in the US.

4.3.2   Wood-Plastic Composites

Wood flour or fiber, mixed at a 50–75% loading with simple thermoplastics, will likely remain the most common way in which builders can claim renewable material content in products (and recycled content too, since much of the wood used in WPCs is post-industrial sawdust). WPCs can be directly extruded from its mixed raw ingredients as decking or even as more complex, hollow PVC-based window profiles, first accomplished in 2007 [26] and now offered by several suppliers using current extrusion technology from Europe, North America, and China.

New processes have also created WPC alternatives to popular PVC siding. England-based Tech-Wood International Ltd. reportedly bases its process on more environmentally friendly polypropylene, impregnating it into long pine wood fiber which is aligned during processing, creating a product that is 75% wood. Recycled polyolefins are popular matrix materials for WPC, but recycled engineering polymers such as ABS result in stiffer WPCs, and could be a potential destination for post-consumer ABS and other plastics reclaimed from scrapped electronic products [27].

Yet there are difficulties in processing and using WPCs. Wood flour easily absorbs water, requiring proper drying before processing, and coupling agents are required to bond the wood to the polymer. Porous WPC products can be subject to mold and mildew, and so processing temperatures must be limited, to avoid causing porosity or degrading the wood. A WPC must be adequately flame retardant, and if used outdoors, contain adequate UV stabilization additives and antioxidants to prevent color change, chalking, and cracking. WPC decking can be slippery to walk on without proper surface texturing, and WPCs are not always maintenance-free as promised, with rotting and splinters possible as with wood [28]. These caveats and others have resulted in consumer complaints (and occasional lawsuits) directed at WPCs producers over the past decade of use – especially when a WPC is touted as eliminating the problems of wood in order to justify its greater cost, and then fails to do so. Still, better processing and handling techniques are allowing this kind of recycled/bio-based material to grow in use.

By contrast, bio-based polymers are not appropriate for construction applications if they are biodegradable (after all, construction products must *resist* degradation over the long term). Still, some bio-based polymers that are non-biodegradable are being included in products, with manufacturers emphasizing the LEED credits that can be gained from using them. For example, Invista offers a nylon carpet fiber with 10% bio-based content made from castor beans, a material choice that can increase a user's LEED rating [29]. Or, as way of adding bio-content to PVC wire and cable coverings, Dow Corning offers a renewable feedstock-based, phthalate-free plasticizer [30]. Still, applications like these are likely less impactful than the complete fossil fuel-to-bio-based plastics changeovers that are being developed in the packaging sector.

4.3.3   Key Points

A few key points are worth noting about plastics and sustainability in building and construction:

1. Unlike trends in other application areas, the construction industry still relies heavily on PVC for siding, window profiles, and other products. Because these PVC products are inexpensive, are not personal-contact applications but rather building materials, and are mostly made from rigid rather than plasticized PVC, consumers may not become as easily sensitized to the material's negative environmental impacts. Still, material suppliers for the industry have done some research into incorporating more recycled PVC content into these products, though the economic obstacles are difficult to overcome.

2. The building and construction industry has made notable efforts to increase the recycling and renewable content in its products. The LEED rating system and partnerships such as CARE aid in motivating sustainable choices and in making these practices more economically compelling.

3. Wood-plastic composites are becoming long-lasting replacements for wood in construction, while also containing recycled waste wood and/or resin content. Otherwise, options are limited in building and construction for plastics based on natural resources, given the extreme durability requirements.

4.4 Automotive Plastics and Sustainability

Compared with building and construction, automotive uses of plastics put their diverse aesthetics and engineering properties on much more obvious display. Over a hundred kilograms of various plastics are now used in the average vehicle – from decorative interior plastics to structural engine components. Nearly all the basic families of polymers are represented in various automotive uses, making it the most diverse plastic consuming sector.

The requirements of automotive plastics range mainly in between those for packaging and construction plastics. Automobiles are not disposable, but neither do they have decades-long service expectations, as in construction. Many automotive plastic parts are in personal contact with the driver, while others have out-of-sight mechanical or structural functions. And although most automotive plastics are recyc*lable*, the infrastructure is almost nonexistent for dismantling, collecting, and separating various plastics parts into their polymer families, and then recycling the plastic.

4.4.1 Fuel-Saving Contributions of Plastics

Despite their limited recycling, automotive plastics do serve a major role in overall sustainability – they make vehicles lighter, improving their fuel efficiency and thus reducing greenhouse gas emissions and fossil fuel use. In a vehicle, a 10% reduction in vehicle mass – from the use of plastics, fiber composites, magnesium, aluminum, and high-strength steel – increases fuel economy by 6–8% [31]. Moreover, every mass reduction has a "decompounding" effect, meaning that a lighter car requires a smaller and lighter engine and a lighter structure [32]. Thus, this effect can lighten the overall vehicle by roughly twice the amount of mass that was removed from just one car component (as long as new features are not added to the car that offset or even cancel out the mass reduction!).

For the next few years at least, government regulations will make this "lightweighting" the key positive contribution to sustainability that automotive plastics can make. The United States announced requirements for increased average vehicle fuel efficiency (Corporate Average Fuel Economy, or CAFÉ standards) to 54.5 miles per gallon by 2025. Lightweighting is said to be one of the more cost-effective ways of reaching this goal, costing manufacturers just $1–$1.50 per pound saved per vehicle. A second phase of cuts requiring over 50% reduced emissions by 2025, would require even more severe lightweighting of 20–30%, made possible by the use of even more plastics and light-weight metal structures in vehicles [33].

But where do plastics' most effective contributions lie in reducing vehicle weight? Resin and components manufacturers are looking at nearly every area of current automotive vehicle design for potential savings, and especially at ways of replacing metals in vehicles. Metal brackets, fasteners, and clips are being replaced by thermoplastic and thermoset materials. The polymeric versions of these small but critical components don't rust, protect other components from corrosion, and are easily removed, all of which reduces assembly time and warranty costs from loose or corroded parts [34]. Engineering polymer grades are replacing metal in structural supports, oil pans, and even in housings, in the next generation of electric vehicles. And plastic fuel tanks, weighing 30% less than steel tanks, are already used in most U.S. vehicles, though their overall emissions-cutting effect is offset by their lack of recycling (unlike steel, which is commonly recycled from shredded vehicles).

Thus there is environmental motivation to use plastics in vehicles. However, automotive plastics still have major progress to make in our two main measures of plastics sustainability: recycling and the use of bio-based materials.

4.4.2   Recycling and Automotive Plastics

Slowly, more and more recycled material is being used in automotive plastic components. Typically, automotive plastics "recycling" has meant simply reincorporating scrap resin (regrind) from molding or forming processes into new parts. But more varied post-industrial scrap resin and even post-consumer material, otherwise destined for landfills, is now being adopted for some auto parts, given the pressures on automotive manufacturers to be greener. However, it is not easy to use recycled resin in this application sector, because auto parts are carefully designed to meet strict engineering requirements. Just a little partially degraded or off-spec recyclate can be enough to push material properties out of their allowable range. Thus the use of post-consumer recycled plastic can be limited to 10% in most applications, with levels above 30% being extreme levels reachable only in noncritical parts.

Recycling automotive parts at the end of vehicle life is an extremely complicated case as well in terms of its practical and economical difficulties. Although auto parts' polymers typically have a high value at the beginning of their service lives, they quickly reduce in value after only 10 years of use. The reasons for this are many. Plastic formulations for various automotive parts use hundreds of combinations of polymer types, grades, and fillers; parts are often painted or colored; they are often bonded to, or contaminated by various materials; and they may have faced years of heat or chemical exposure that has degraded their properties permanently. Even with better recycling methods, the automotive plastics recycling "loop" is difficult to close. Despite years of targeting automotive plastics for recycling, little major progress has been made in actually using recovered auto plastics directly in new auto applications. One approach is to focus on recovering larger plastic parts made from one resin grade. Large parts can be dismantled from vehicles as a whole before vehicle crushing and shredding, and their high-value resin grade can then be identified for recycling, typically by referring to codes molded into the parts or by other methods.

Organized efforts are targeting the practical recovery of plastic auto parts, and recycling automotive vehicle waste in general. For example, the European Union (EU) has set high targets for vehicle recycling overall with its End of Life Vehicles (ELV) Directive. The year 2000 directive called for 80% of discarded vehicles to be recycled or reused by the beginning of 2006, and 85% by the beginning of 2015 [35]. The Most EU countries appear to be complying with the 2006 target thus far, although the largest share by far of recovered vehicle materials is composed of metal parts. This particular directive was modified in 2014 and 2018 to address major challenges related to illegal shipments of ELVs, highlighting ongoing complexities of managing a supply chain across an entire continent. Still, an evaluation of the ELV directive in February 2020 suggested that plastics recycling remains an important component of the overall effort [36].

Recycling large plastic parts typically requires more than just throwing them into a shredder for size reduction. Parts such as plastic bumper fascia are painted, and the paint acts as a contaminant in the recyclate, ruining the surface quality of new fascia. This recycled fascia material can be more readily used in similar but less aesthetic exterior auto applications such as underbody deflectors – though this does not ideally close the loop on their recycling. Plastic parts recycling can also be complicated by adhesive tapes or foam stuck on the part, by molded-in or embedded metal fasteners, or by other integrated parts. Design principles (see Chapter 5) can prevent these obstacles to recycling from becoming a problem before the mold for creating a new automotive plastic part is even cut.

There are other major end-of-life collection and separation issue for automotive plastics. The plastics-containing nonmetal "fluff" mixture that remains after vehicle shredding is generally only recyclable using complex chemical processes that reduce all plastics to simpler molecular forms for reuse. However, there may be some hope. New processes in mechanical and sink/float separation of polymers in shredder residue have been commercialized after pioneering efforts by MBA Polymers and others over the past decade. New methods and techniques, first proven at Argonne National Labs in the 1990s, can separate up to 95% of shredder residue polymers by their densities and by their hydrophobic or hydrophilic natures [37].

4.4.3   Bioplastics in the Automotive Industry

As in building and construction applications, automotive parts require resilient, durable, long-term plastic-based materials. They must be unaffected by water, light, impact, and extremes of temperature, thus eliminating most biologically based polymers that would be considered good fits in the packaging sector. Nonetheless, automobiles are shorter-term applications than buildings, and are seen by the public as somewhat disposable, considering the constant stream of new models auto manufacturers present us with. This partial disposability puts pressure on manufacturers to use more natural, renewable materials.

To increase plastics' bio-content, automotive manufacturers have mainly turned to exploiting the engineering properties of light-weight natural fibers for reinforcing plastics. Henry Ford had made some early efforts in using hemp fiber in cars, and recently these plant fibers have again drawn interest. Automakers are studying how to standardize the use of these fibers in higher-volume vehicles for a variety of components.

Natural fibers for automotive plastics composites come in different categories, from nonwoven, short wood fibers, to partially woven mats made with non-woody long plant fibers. Flax, hemp, wheat straw, and abaca (banana) fibers are just a few of the plant sources (see Chapter 3). Natural fiber-filled plastics are already being used in several interior auto applications, such as door panels, trunk liners, glove boxes, seat backs, and roof frames, such as that used in the Mercedes Benz E-Class [38]. In European passenger cars alone, 25,000 tons of these fibers are being used per year, with new plastic composite applications being developed for them all the time, including the 3rd generation of Porsche's "BioConcept Car", the first mass-produced car to feature biofiber composite body panels [39]. And manufacturers are not shy about making these fibers visible in components, thus making the bio-content of the vehicle obvious to consumers. By allowing natural fibers to be visible at the surface of the parts, they are advertising their use.

Such explicit attention to the fiber reinforcement in composites was once reserved for carbon fiber, which is perhaps the most effective reinforcement for creating low-weight automotive structures. Though it cannot be called a natural fiber, new production processes make carbon fiber from natural precursor materials such as cellulosic materials and lignin (such as kraft paper pulp), and even from recycled polymers. In terms of strength in lightweight automotive polymer composites, carbon fiber beats natural fiber easily; in terms of cost, the overall case for carbon fiber use in standard automobiles remains hard to make. Experts suggest that improved manufacturing could reduce final carbon fiber costs by 35%, where it would be competitive enough to be used across many automotive applications [40].

Meanwhile, the newest biopolymer resins themselves are being stiffened and made more heat resistant for automotive use. Resin producers such as DuPont have also created partially bio-based engineering grades that approximate the properties of engineering resins commonly used in automotive.

Traditional polymers based on renewable raw materials also interest automakers. Polyethylene and polypropylene made from renewably sourced ethanol will let some automakers take credit for green design. The inclusion of more bio-based plastics may depend on the use of these bio-based traditional resins, which auto designers already have great familiarity with and confidence in.

4.4.4   Key Points

The automotive industry was under stress during the recent Great Recession, functioning in survival mode. Still certain trends related to plastics sustainability are evident, coming from the industry's response to environmental and regulatory pressures:

1. The continued lightweighting of vehicles may be a decades-long quest in response to higher fuel mileage requirements, calls for lower greenhouse gas emissions, and the potential for steadily increasing oil prices. As the lightest-weight engineering materials, plastics and plastic/fiber composites will showcase their most positive environmental talents by playing lead roles in all areas of future automobile designs.

2. The recycling of automotive plastics continues to be limited by many factors, both practical and economic. Much higher virgin resin prices, plus governmental pressure, may be required before auto plastics are recycled at higher levels. Conversely, the incorporation of recycled plastics back into automotive parts is difficult because of the strict property standards needed for a durable engineering application.

3. Bio-based plastics for the automotive sector will gradually include more natural-based fibers as reinforcements for polymers. Bio-based polymer resins may be limited to special partially bio-sourced traditional polymers that have the necessary engineering properties.

4.5   Specialized Applications and Plastics Sustainability

The three application areas above – packaging, construction, and automotive – show the wide range of ways in which plastics are used, and in which their sustainability is being tested. But the few smaller-volume application sectors covered below have particularly interesting sustainability issues of their own. Some of these specialized applications are inherently incompatible to recycling, recycled material content, or bio-based content; others are not, or have not had their life-cycle impacts sufficiently studied to determine sustainable options.

4.5.1   Electrical/Electronics Applications

Plastics used in consumer electronics seem to be mostly invisible to consumers, who are now easily distracted by the increasing functionality of smart phones, portable computers, and similar devices. But most of these devices are housed in plastic that must be durable over a lifetime of use (which now may be only a few years or less, before the "old" device is disposed of and replaced with newer technology). In this way, electronics plastics have some similarities with interior automotive plastics – they must resist the impacts of use, and maintain an aesthetic surface finish. Unfortunately, also like automotive plastics, their recycling infrastructure is limited, and bioplastics are only marginally used in the sector.

But opportunities (and markets) do exist for using bioplastics in these devices. Cell phone casings are one high-volume application that could see major life-cycle impact reductions from bio-plastics, given how often consumers replace their cell phones, and how rarely the phones are recycled. For example, PLA reinforced with kenaf fibers was developed as far back as 2006 as a casing material by NEC Corporation of America and UNITIKA Ltd. Loaded at 20%, the natural fiber reportedly improved the PLA's heat distortion temperature by about 60°C (up to about 120°C), and roughly doubled its modulus of elasticity and impact resistance, making it suitable for the application [41].

Other green points may be gained by manufacturers being required to accept old electronic devices for recycling. There is obvious interest in reclaiming their engineering thermoplastics. These "take-back" requirements in European Union countries may have had questionable success thus far, though voluntary electronics recycling events held in the United States draw thousands of consumers seemingly eager to get rid of tons of old electronics in an environmentally responsible way.

However, recycling operations for electronics expose workers and the environment to potentially hazardous materials. For example, electronic device materials often contain halogen-based flame retardants which have questionable health effects and are persistent in the environment. Recycling operations in developing countries have been found to be particularly lacking in proper controls for worker protection. Documentaries such as "Plastic China" (2017) and new regulations for polybrominated diphenyl ether flame retardants found in some electronic equipment have exposed the dangers of unregulated waste management [42].

And so the recycling of electronics plastics is no quick fix for enhancing sustainability. However, the trend toward "convergence" in electronic devices, allowing a single device to have a growing number of functions that would previously require multiple devices, may at least reduce the overall total of discarded electronics plastics.

4.5.2   Medical Plastics and Packaging

The impact of COVID-19 has led to renewed interest in and growth of plastic personal protective equipment (PPE). New market studies suggest growth of 17% through 2021, resulting in a $29 billion market [43]. Plastic applications in the medical and healthcare arena range from packaging that is similar in nature to food and beverage packaging to devices used inside the human body, and from electronic equipment housings to various textiles and accessories unique to hospitals and clinics. Pharmaceutical containers, syringes, intravenous (IV) bags, and trays, most of which must be sterilized, are just a few examples of medical plastics uses which are consuming plastics at a steadily increasing rate, even before the pandemic. Polypropylene is the leading resin for these products, followed by low-cost PVC (the use of which in healthcare appears to be discouraged not quite as much by its negative image as it is in other PVC application sectors).

But given the nature of their business, healthcare professionals are generally sensitive to the compositions of plastic products. Obviously, they wish to avoid any materials with the potential of negatively interacting with the human body or treatment. Thus, more professionals are noting whether a flexible PVC has phthalate-based plasticizers in it, for example; some are considering the possible influence of bisphenol-A in polycarbonate products. In these special cases, healthcare professionals might seek materials such as phthalate-free PVC medical tubing, or Eastman Tritan™ copolyester (a non-BPA alternative to PC), respectively. Antimicrobial plastic additives can be manufactured into many polymers now, with companies such as Microban offering powder and liquid-based masterbatches for a variety of plastic processes [44].

As anyone who has visited a modern hospital already knows, most medical products are for one-time-use applications. This seems to be true especially for plastic products, which often end up being sent to hospital incinerators or other permanent disposal routes, rather than being recycled. Contamination from medication and bodily fluids means that significant purification steps would be required during recycling operations for medical plastic waste streams. Likewise, the threat of cross-contamination also means that recycled plastic is not often used in medical devices [45]. And as with all medical applications, products containing recyclate or based on renewable resources would need absolutely perfect traceability to verify the purity of their materials and processing. This is a challenge even when traditional virgin medical resin grades are used.

So in general, sustainability issues are not priorities for health-care product manufacturers and users. Even if their customers were interested in greener products, medical manufacturers are slow to make changes that would reduce their environmental footprints. Rather, functionality and cost-effectiveness are their key concerns. For example, biologically derived resins are used for implanted products because of the way they interact with the human body, not because they are bio-based; using them for other products would be cost-prohibitive, and could create contamination issues related to their biodegradability. Similarly, material reduction is not often a key issue; a medical tool might be designed with excess material thickness to ensure there is no chance of it failing during use, rather than the minimum thickness required.

The throw-away culture of healthcare is a way of ensuring the germ-free treatment of patients, but it makes it difficult to reduce the environmental footprint of medical plastics in ways other than with design modifications (see Chapter 5). Products designed to be cleaned and reused rather than simply disposed of can be made compatible with a healthcare facility's priorities, procedures, or costing practices. Conversely, any cost premium for using recycled or recyclable or bio-based materials in products does not support efforts to minimize costs in an already high-cost sector of the economy.

It is an uphill climb for those interested in reducing plastic life-cycle impacts in this sector, but things may be changing. The topic of sustainability is popping up in more medical product conferences in recent years, showing a growing interest. The Healthcare Plastics Recycling Council, founded in 2010, has developed a guide for recycling that connects recyclers, hospitals, and other providers to recover valuable materials that meet FDA requirements.

4.5.3   Agricultural Applications

The general public mostly is unaware of the extent to which plastics are used in agricultural production. They are used not only for plant containers, but also for covering fields and as greenhouse films – holding in moisture, spurring young plant growth, and protecting growing crops from extreme conditions and pests. The sector consumes large quantities of the simplest commodity plastics (PE, PP, and PS) for products such as mulch and silage film, bale wrap and twine, landscape edging, plant trays and pots, and pesticide and fertilizer containers and bags. Unlike in medical applications, agricultural "ag" uses have more tolerance for material contamination and property variations – opening up the potential for recycling the plastics and using them again in a similar ag application, or for using plastics that readily biodegrade.

In terms of recyclability, agricultural and horticultural plastic applications do present challenges, considering the dirt and ag chemical contamination they encounter. One of the most contaminated waste streams is PE mulch film, which is collected from fields and then commonly landfilled or incinerated, though some companies like Revolution (Little Rock, AR), have developed recycling business models that include pick up/drop off locations for contaminated mulch films. Despite the contamination challenge, some businesses recycle this material into thermoformed horticultural products such as plant containers. These products can tolerate a relatively high level of contamination, somewhat "closing the loop" for this material by reusing it in the same business sector.

Biodegradable plastics present interesting possibilities for ag uses. They are being studied as mulch films that can simply be rototilled into the field after the growing season and left to degrade, saving on film collection and disposal costs. Bio-copolymers from the PHA family, which are degraded by organisms in soil, have been evaluated as mulch films. Studies have focused on how the water content in the soil and other factors influence biodegradation. Here, researchers found that soil moisture content, up to an optimal level, strongly increases the rate of weight loss of

PHA film. This helps nearly all of the film to degrade within a month or two, while without moisture, there is little or no biodegradation. As developers acquire more understanding about PHA soil biodegradability, other studies are needed for comparing the costs of using this film with the costs of using, collecting, and then disposing standard PE mulch films.

4.6   Conclusions about Sustainable Plastics Applications

This was just a brief overview of sustainability-related plastics applications. It can only assist in developing a comprehensive plan or thought process for optimizing new, more sustainable applications, or for selecting materials with reduced environmental impact. Before a material selection process can begin (Chapter 6), we need at least a discussion (Chapter 5) about how the design of various plastics products influences their environmental footprint. Looking at applications is helpful, but it provides an incomplete picture because there are always many "known unknowns" about the real impacts of specific products. Even well after a product is released, questions can remain about whether it fulfilled its intended sustainability goals. For example, if a product is designed for easy recyclability, are consumers actually recycling it? Or if it is biodegradable, has this property brought any real gains to the environment, or is the product simply being landfilled along with other plastics? Or is it being littered more because consumers believe it will simply degrade anywhere. Or is it acting as a contaminant in the recycling stream?

Another issue concerns heightened expectations. Sometimes green plastic applications have been reported on with perhaps excessive optimism and a lack of objectivity. In reality, some are just too new to be judged in terms of long-term sustainability. And for many of these new sustainable applications, the consumer or market response is not yet clear.

Still, some general trends and conclusions can be proposed about plastic applications and sustainability:

1. Manufacturers and marketers are identifying market niches where plastics with environmentally sustainable benefits have a chance to compete with traditional fossil-fuel-based plastics.
2. Certain application sectors (e.g., construction) already use significant amounts of plastics based on renewable or recycled content.

3. The big picture advantages of all kinds of plastics in certain applications (for example, helping to save fuel consumption in automotive transportation) may be more important than environmental impacts related to their production, recycling, or renewable content.

4. Some application sectors may remain inherently unfriendly to or apathetic about recycled or biodegradable plastics (e.g., healthcare), while others (e.g., packaging and agriculture/ horticulture) may become more friendly to these kinds of materials as they are developed.

5. The current best-choice plastic formulation that would be considered environmentally friendly in nearly every application sector would be based on recyclable, traditional, simple polymers (such as the polyolefins PE and PP) that are based on renewable feedstock sources. (This assumes the biomass used for these feedstocks is grown without causing excessive environmental damage and does not interfere with food production).

References

1. Carlin, C., Global Dispatches: 6th Annual AMI Thin Wall Packaging Conference North America", Conference Report. *Plast. Eng.*, 73, 8, 30–34 2017.
2. Global Packaging Landscape: Growth, Trends & Innovations. *Euromonitor Int.*, 2019 July.
3. Loepp, D., Hypocrisy in 'Plastics-Free February, in: *Plastics News*, http://www.plasticsnews.com, 2011, February 7.
4. Carlin, C., AMI's Thin Wall Packaging Conference Offers a Look at Germany's Innovative Circular Economy Strategies", Conference Report. *Plast. Eng.*, 75, 2, pp. 6–11, 2019, February 5.
5. ABC News, https://abcnews.go.com/US/plastic-bag-bans-helping-environment-results, 2020, February 23.
6. More Recycling report. *2017 National Post-Consumer Plastic Bag & Film Recycling Report*, July 2019.
7. https://www.recyclingtoday.com/article/the-economics-of-pet-recycling/
8. www.nextloopp.com
9. Plastics: The Facts 2018: An Analysis of European Plastics Production, Demand, and Waste Data, in: *PlasticsEurope*, Trade Association.
10. Rethinking the Paper Cup, in: *Nicole Whitemann, NatureWorks, presentation at ITR*, 2018.

11. Carlin, C., Innovation Takes Root in California. *Plast. Eng.*, 74, 10, 8–13, 2018, December 5.

12. Ibid.

13. Hamad, K., Kaseem, M., Ayyoob, M., Joo, J., Deri, F., Polylactic acid blends: The future of green, light and tough. *Prog. Polym. Sci.*, 85, Oct. 2018, 83–127, 2018.

14. https://www.nytimes.com/2020/02/10/business/dart-foam-recycling.html

15. Chauvet, M., Sauceau, M., Baillon, F., Fages, J., Mastering the Structure of PLA Foams Made with Extrusion Assisted by Supercritical CO_2. *J. Appl. Polym. Sci.*, 134, 2017, March.

16. Tolinski, M., Building new opportunities for plastics. *Plast. Eng.*, 64, 9, 6–8, 2008, October.

17. Vinyl Institute, www.vinylinfo.org/uses/building-and-construction, July 2020.

18. Mordor Intelligence, Wood Plastic Composites Market Growth, Trends, & Forecast, 2020-2025.

19. Ward, S.K., Green mandates: Opening up a new sales channel and growth potential for materials companies, ANTEC 2011 [Proceedings]. *Soc. Plast. Eng.*, 2011.

20. https://www.epa.gov/greenerproducts/recommendations-specifications-standards-and-ecolabels-federal-purchasing

21. U.S. Green Building Council; LEED Online, 2020.

22. Altshuler, K., Horst, S., Malin, N., Norris, G., Nishioka, Y., *Assessment for the technical basis for a PVC-related materials credit for LEED (Report)*, U.S. Green Building Council, 2007, February.

23. Tarnell Company VI Survey, *Sold Amounts*, 2017.

24. Tolinski, M., Making the "unrecyclable" recyclable. *Plast. Eng.*, 65, 10, 6–7, 2009, November/December.

25. Carpet America Recovery Effort, *Annual Reports*, 2009/2019.

26. Schut, J.H., First direct extrusion of complex PVC window profiles. *Plast. Technol.*, 2007, May.

27. Lauzon, M., Researcher studies composite using wood and recycled ABS. http://www.plasticsnews.com, 2010, May 6.

28. Sidler, S., The Problems With Composite Decking, company blog https://the-craftsmanblog.com/the-problems-with-composite-decking/, 2018, April.

29. Bio_AntronTM carpet fiber, http://antron.net/content/product_line/ant24_12.shtml, 2011.

30. Dow introduces first bio-based plasticizers for wire and cable applications. Press Release. http://www.dow.com/wire/presscenter/press_release/2010/20100412a.htm, 2010, April 12.

31. www.automotiveplastics.com, July 2020

32. Malnati, P., Automotive Composites: Mass Reduction for Mass Production. *Plast. Eng.*, 72, 8, 2016, September.

33. https://www.assemblymag.com/articles/94341-lightweighting-is-top-priority-for-automotive-industry, 2018.

34. Malnati, P., Polymeric Fasteners Gain Traction in Automotive. *Plast. Eng.*, 76, 5, 24−27 2020, May.

35. End of life vehicles, Europa Summaries of EU Legislation, http://europa.eu/legislation_summaries/environment/waste_management/l21225_en.htm, 2011, March 5.

36. ELV Evaluation Workshop, Meeting Minutes, https://www.elv-evaluation.eu/fileadmin/user_upload/documents/ELV_Evaluation_Workshop_Minutes.pdf, 2020, February.

37. Toto, D., Re|Focus 2016: Recovering Plastics from ASR, Recycling Today, 2016, May.

38. Caliendo, H., Natural Fiber Composites Gaining Traction in Automotive, www.compositesworld.com, 2016, August.

39. Winkelmann, N., *Porsche Production Car Produced Using Natural Fiber Composites*, Composites World/LightweightDesign Magazine, 2019 September.

40. Nunna, N., Blanchard, P., Buckmaster, D., Davis, S., Naebe, M., Development of a cost model for the production of carbon fibers. *Heliyon*, 5, 10, 2019. via Science Direct.

41. Complete mobile phone housing made of PLA, reinforced with kenaf fibres. *Bioplastics*, 1−1, 18−19, 2006.

42. Zenneg, M. *et al.*, Formation of PBDD/F from PBDE in Electronic Waste in Recycling Processes and Under Simulated Extruding Conditions. *Chemosphere*, 116, 34−39, 2014, December. (via ScienceDirect).

43. Esposito, F., Study Expects Medical Plastics Growth to Top 17%, Plastics News.com, 2020, April 28.

44. https://www.microban.com/antimicrobial-solutions/applications/antimicrobial-plastics

45. Esposito, F., Sustainability an issue with medical product design, in: *Plastics News*, http://www.plasticsnews.com, 2010, April 20.

5

Design Guidelines for Sustainability

The design of plastic products has a major impact on how much material they consume and how efficiently they fulfill consumer needs. Designs need to be optimized for both the material's capabilities and product requirements – as well as for making the products recyclable or reusable. Although this chapter is not a complete guide on the principles of plastics design, it does cover design considerations that influence the environmental impact of products. (The next chapter will incorporate these considerations with other property and life-cycle issues into a discussion about the material-selection process.) The structure of Chapter 5 is simple:

- Six green design principles and guidelines for plastic parts – these will be put into context, showing how they help meet sustainability goals during the design phase (5.1).
- Consumer preferences – the "wildcard" in green design (5.2).

Michael Tolinski and Conor P. Carlin. Plastics and Sustainability 2nd Edition: Grey is the New Green: Exploring the Nuances and Complexities of Modern Plastics, (159–174) © 2021 Scrivener Publishing LLC

Starting in the 1990s, organizations such as the American Plastics Council distributed publications providing guidelines on Designing for the Environment (DFE), Designing for Recycling (DFR), and related topics. Many of these publications' guidelines were simply good business practice because they focused on ways of using material efficiently in a part, such as using the thinnest possible wall or gauge thickness. Other DFE/DFR guidelines concerned designing for disassembly (so that parts can be separated for recycling), designing molded-in fastening features as alternatives to separate fasteners or adhesives, and designing molded-in decorative elements as a way of avoiding external coatings and finishes, which increase plant emissions and make the part less recyclable. These remain sound and usually cost-effective principles, and will be touched on below.

But with greater social consciousness of environmental issues, manufacturers and retailers are revisiting these principles. Sometimes they take them to extremes, willing to sacrifice certain features or properties to reduce packaging material. Some companies intentionally make their product changes obvious to the consumer to gain green credibility, if not for inherent environmental benefits. In the era of the circular economy, more effort is being put into material selection, including the development of mono-material packages or dual-material packages that can be easily separated by the consumer for recycling. Bonus/malus schemes in Europe reward or punish companies according to material and design criteria [1].

This kind of marketed DFE is becoming more common, as designers, manufacturers, and retailers realize there may be green points to score by optimizing designs for minimal environmental impact. Moreover, packaging experts agree that the design stage is where the biggest sustainability gains can be made. But many also say that the era may be ending for making big improvements by redesigning packaging with thinner walls or reduced material usage. There is indeed a limit to how thin a PET bottle can be and still function, for example, and this limit is being approached.

Thus, new directions are required for more sustainable plastic parts. New designs may be needed to accommodate the properties of post-consumer recycled, bio-based, and/or biodegradable materials. Meanwhile, the consumer may need to be educated and persuaded more to recognize and accept the product changes in greener packaging, especially if there are trade-offs in functionality. The new directions for green design of plastic products need to be re-envisioned in this context for contemporary decision making in the real marketplace.

5.1   Green Design Principles

In Chapters 2 and 3, the "Twelve Principles of Green Chemistry" helped guide the discussion about evaluating the sustainability of plastic materials. In this chapter, Anastas and Zimmerman's related "Twelve Principles of Green Engineering" from their 2006 book [2] include useful and relevant guidelines for plastics green design, such as the following selections:

- "System components should be designed to maximize mass, energy, or temporal efficiency."
- "Embedded entropy and complexity must be viewed as an investment when making design choices on recycle, reuse, or beneficial disposition."
- "Targeted durability, not immortality, should be a design goal."
- "Design for unnecessary capacity or capability should be considered a design flaw. This includes 'one size fits all' solutions."
- "Multi-component products should strive for material unification to promote disassembly and value retention – (minimize material diversity)."
- "Design should be based on renewable and readily available inputs throughout the life-cycle."

Paraphrasing each of the above guidelines in a way that is more pertinent to plastics product design, we propose the following versions of these statements:

- Product designs should require as little plastic as possible while still allowing the product to function and be aesthetically acceptable.
- Higher-value, more complex plastics should only be used when required, and the design should fully exploit their special properties. The plastic materials should also be recoverable for recycling, according to their value.
- Plastic products should be designed only to be durable enough for their expected use-lives and situations of use.
- The use of excess plastic in a product for reasons other than minimum functionality should always be questioned.

And plastics products should be customized for their required function.

- Single-material plastic products are preferable so that the manufacturing process is simple and efficient and the product retains the practical value of being a pure, uncontaminated material, making recycling simpler and cost-effective. (This principle also means that diverse additives in plastics are generally unwanted, when their function can be integrated into the polymer or design itself.)
- Designs should allow the use of materials that are based on efficiently made renewable feedstocks.

As Anastas and Zimmerman affirm about their principles, they can sometimes come into conflict during the design process, and choices must be made for prioritizing one over the other (one might even be able to imagine plastic part designs in which practically all of them come into conflict). And by sticking too religiously to the principles, a "green" plastic parts designer might ignore alternative approaches or materials – some of which may actually be better options. Focusing on one narrow set of options or design concepts may limit the designer's abilities to make optimally sustainable choices. Or sometimes a lack of attention may be given to long-term life-cycle issues with a product or material, resulting in an overall less desirable design. For example, a part might be designed to use a biodegradable material that ends up contaminating the overall recycling stream for the product. (This is why any DFE or green design approach should be incorporated within the overall material selection process, as discussed in Chapter 6.)

The principles do not focus only on recyclability or the use of bio-based materials. Rather, a common theme behind many of the green principles is simply: "Do not overdesign." Even though alternative materials have been presented in this book thus far, most green design situations are far too complex to have a simple answer such as: "Design to use more recycled material." In fact, with numerous multinational brands pledging to increase their use of recycled content, supply-side economic issues have come to the fore [3]. Thus, this chapter and the remainder of this book will view bio-based or recycled materials simply as options for addressing principles of green design and minimizing life-cycle impacts in material selection.

Each subsection below focuses on one of the above restated principles for green plastic product design. The headings have turned the wordy principles into more memorable guidelines.

5.1.1   Minimize Material Content

Product designs must of course always be functional and have a good or acceptable appearance, while requiring as little plastic as possible. Apart from the benefits of a reduced environmental footprint, minimizing material content is simply good business practice that contributes to the bottom line. For common plastics, this kind of lightweighting usually depends on reducing the thickness of a product, and/or using a less dense plastic. In a rigid container, stiffening features such as ribs may allow wall thickness to be reduced. Or a container might be designed to accommodate a stiff but low-density plastic like polypropylene, rather than PET, for example. In engineering applications, thin-wall molding allows weight reduction of the entire engineered system, multiplying its benefits, especially if the system is weight sensitive, like an automobile, aircraft, or portable electronic device.

Lightweighting/thinwalling efforts have been most obvious in beverage bottle production. Over the past 10-15 years, standard PET bottles (0.5L or 16oz versions) have become thinner and lighter, shedding almost 50% of their weight over a ten year period [4]. Getting closer to a theoretical limit of PET bottle thickness, designers are looking to add structural features that aid in the lightweighting of bottles. Sometimes a variant shape or feature can be designed that makes a bottle unique, while also reducing its environmental footprint.

Another, less often discussed approach for reducing mass in a design is to radically reduce the density of the plastic by foaming it, giving it a cellular structure. Foaming allows for structures with relatively thick and stiff walls that are lighter overall when compared to a structure of equivalent stiffness using a solid resin. Most thermoplastics can be foamed. The molding process can create a "sandwich" structure having a foamed core and solid resin skin, resulting in a material with a high stiffness-to-weight ratio and a good appearance. An alternative like this may become important if the emphasis on material reduction shifts more from the packaging arena to engineered materials. Economics come in to play again, however, as recyclers are forced to adapt to changing part weights and volumes. In commodities markets, plastic is sold by weight, not volume. The lightweight bottles of today, by definition, use less material than their predecessors from a decade ago, therefore more of them is required to reach the same weight threshold. Because costs associated with recycling are based on volume, but revenues are measured by weight, the overall economics appear to be structurally at odds with design for DFE.

Such emphasis on lower weight parts with thinner part walls may at times dictate that a stronger, higher-value engineering or reinforced plastic is more appropriate to use than a cheaper plastic. However, in these cases, the designer must evaluate and compare the system costs of the stronger material, its processing characteristics, its recyclability and other life-cycle impacts, and the degree to which the new design/material satisfies (or does not satisfy) the other green design guidelines below. For example, the higher-value plastic may make a relatively simple part overdesigned, having certain extra engineering characteristics (besides strength/stiffness) that are not at all required for the application.

5.1.2   Exploit a Material's Full Value

This guideline further develops the preceding idea that the sustainability of a design can be optimized using a higher-value plastic having higher properties, such as an engineering resin instead of a commodity resin. A stronger plastic may allow a part to survive through its entire service life without failing and having to be replaced. A stiffer plastic may allow the walls of a part to be thinner, contributing to the goals of the lightweighting discussed above. Or the higher-value material may simply be the only option that allows the design to work. Overall green principles dictate that more complex plastics should only be used when required. And no matter what resin grade is used, an efficient, optimally sustainable design should fully exploit its properties.

The full value of a material can also be exploited by reusing plastic that has already been used in a product. Incorporating high recycled content (especially post-consumer content) may require using resins with a broader range of property variations than virgin resins, due to partial degradation of the polymer that occurs during its previous processing and use. This might even require designing a more generous wall thickness, for example, to ensure adequate part strength or stiffness. The design team would need to choose whether the positive benefits of using recycled material outweigh the potential costs of greater part thickness and mass or the greater risk of part failure. They should also consult with manufacturing engineering about processing recycled-content resin to see if the resin's melt-flow properties are acceptable and consistent enough. Here, following a concurrent engineering process that links design and manufacturing is a great help in designing for sustainability.

Whether using recycled content material or not, the value of the plastic product and/or its material should also be exploitable when the product's service life is over. Thus the design should allow the product to be recoverable for reuse or recycling. The principle also implies that the higher the value of the material used in a design, the greater the efforts should be on the designer's part to ensure that the plastic part can be recovered from the waste stream. (Ironically, this principle mostly conflicts with the real world, in which it is the high-volume, low-value commodity plastics that are more commonly recycled.)

To encourage product recycling or reuse, designers can design component products for disassembly. Fastening points or hinges integrated into the part should allow it to be easily and cleanly separated from its assembly, especially if the component can be reused as a whole. The part's material identification code should be easily readable on the part, whether it is a number code associated with determining its recyclability, or a clearer code explaining the part's composition, like the ones used on plastic automotive parts.

But before being disassembled or discarded for recycling, much value can remain even in many commodity plastic products. The direct reuse of plastic packaging products seems as if it would be rare, but there are common examples. Plastic shopping bags are reused as trash can liners or as receptacles for picking up after pets – for these reuses, most "T-shirt" bag designs are nearly optimized. Reclosable rigid food containers are often washed and reused for food storage; chemical containers become water-hauling buckets; and scrap yards are inexpensive sources of reusable auto parts – many of them made from high-value engineering plastics.

Because plastic products often possess this kind of durable reuse potential, innovative designs can actually encourage their reuse. Attractive, durable water bottles designed for refilling encourage the consumer to choose refillable bottles over single-use bottles (which themselves are often refilled several times before being discarded/recycled). But this is just a simple example. In recent years, refillables have gained popularity with refill stations popping up in many stores beyond just natural food markets. The Zero Waste Movement has sparked the growth of a niche industry where conscientious shoppers can buy in bulk using refillable containers while avoiding single-serve, disposable plastic containers. Yet even here, where glass containers are perceived as "more" reusable than plastic, well-designed plastic containers prove that they can be washed and reused many times without posing the same breakage hazards associated with glass. The launch of TerraCycle's Loop platform in 2019 pushed the concept further, bringing convenience and environmental peace of mind to urbanites with disposable income [5].

A couple of final examples can be used to illustrate the idea of exploiting a plastic's full value in a part. Because properly designed plastic parts in nearly any material can be tough and durable enough for repeated use, reusable plastic containers (RPCs) for shipping are a great case example in minimizing total environmental footprints. Each time an RPC is used for transporting products such as food produce, it eliminates the need for a disposable form of packaging to be created and used for the purpose – avoiding the environmental impacts of single-use shipping packaging. Life-cycle studies have shown RPCs to have lower long-term impacts than disposable corrugated packaging; researchers have calculated that RPCs contribute 39% energy savings and 95% less solid waste through their total life cycle [6]. Moreover, RPCs are lightweight and cleverly designed, saving fuel during transportation and making them easy to handle and stack.

Reusability can even be designed into a high-end application that uses plastic components, such as a complex electronic surgical tool. A green redesign here can minimize the mass of the system's single-use disposable components while allowing a certain number of reuses of its more durable, higher-value components, rather than disposing of all components with each use.

5.1.3 Fulfill Durability Requirements

In the early years of plastics use, materials and designs often did not survive their products' service use lives. This contributed to plastic products' initial reputation of being "cheap". Perhaps in reaction to this, resin quality quickly improved and designers paid extra attention to making durable products, perhaps even overcompensating for design flaws, as they learned the tricks of plastic part design. Now, if anything, we may be in a world in which plastic products are over-designed for many of their applications, being discarded with much "life" still left in them.

Under this principle, biodegradable polymers would seem to be advantageous, especially in packaging or short-term disposable applications. Biodegradable plastics seem to suit products that are supposed to be used briefly before being disposed of (and usually are not recycled), such as flexible packaging. A degradable plastic material, such as thermoplastic starch, might stay usable just long enough to gain its full value, before it shows signs of wear, degradation, or water damage. (For disposable tableware, this could be a period of minutes, or it could be measured in months for common plastics that contain special biodegradability-enhancing additives.) However, some biodegradable plastics can be costly or require careful processing and handling. And even more importantly, their biodegradability should not be confused with an over-all lack of durability – it is only one aspect that may limit durability. Rather, this principle focuses more on what material properties the product requires that would allow it to be used through its entire use life (plus an engineering margin of safety),

From Design for Environment to Safer Choice: How the US Environmental Protection Agency Denotes Sustainable Products

In the early 1990s, the US EPA developed a non-regulatory initiative that would provide companies with guidance surrounding the impacts on human health that some products might have. Initially labeled, "Design for Environment" (abbreviated to DfE), the EPA worked with a wide array of industry groups, NGOs, environmental organizations, and academia to create objective criteria and technical resources such as Life-Cycle Assessments (LCAs) and Comparative Technology Substitutes Assessments (CTSAs).

Early efforts focused on responding to public awareness about harmful effects of some chemicals. Later, a certification and labeling program was introduced to allow companies that met certain criteria to differentiate their products on store shelves.

In 2015, the program officially changed from "DfE" to "Safer Choice". The logo was also changed and modernized, reflecting a consumer-facing approach. Those manufacturers who want their products to carry the Safer Choice label must adhere to the criteria for safer chemical ingredients and meet requirements for performance, packaging, pH levels, and volatile organic compounds (VOCs).

(Source: www.epa.gov/saferchoice)

but no longer. The principle ensures no extra features or materials are used in the design – so a special mechanical hinge would not make sense when a molded-in living hinge would do the job for a container. Plastics are not infinitely flexible, and living hinges will eventually fail, but if only a finite amount of living hinge flexing is needed over a limited lifetime of use, the product can be designed accordingly.

Plastics have other durability limits to design for. For instance, plastic pipe is stressed by pressure and threatened by environmental stress cracking, meaning that the pipe wall ideally should only be as thick and strong as needed to last until the pipe's normally scheduled replacement. Walls any thicker than an engineering degree of safety would violate the green guideline. In other words, any extra material or features for an excessively durable product create unnecessary environmental impacts, energies, and wastes associated with the product's production.

5.1.4   Minimize Non-Functional Features

Slightly different from those mentioned above, this guideline argues against the use of plastic or other materials in a product for reasons other than meeting the product's minimum functionality requirements. Rather, a plastic product design should be customized as much as possible for its required function, even if it is manufactured in the millions. (And here, aesthetics, which are so important with many plastic parts, could be categorized as a kind of function.) In engineered applications, this guideline obviously supports notions of "elegant design," because it advocates sleek, thin, contoured shapes with minimum features. In many cases, basic customer interviews and usability testing, plus finite element analysis computer simulation, may reveal when a part design has unnecessary features, or the initial design has not been optimized. Here, the sales and marketing department could assist in determining what features the customers/consumers really want in a product, and which ones are dispensable. A customer of a packaging producer may even be willing to agree to the elimination of a desired featured if it can be justified in the name of sustainability.

5.1.5   Focus on Single-Material Designs

Using multiple materials in a design adds complexity and potential waste to a manufacturing process. It also makes a part harder (or impossible) to recycle. Thus, single-material product design, properly done, results in a potentially smaller environmental footprint. Plastics forming processes and manufacturing procedures are particularly friendly to single-material designs. These processes themselves are based on relatively simple concepts: melting or heating a raw plastic material and then molding or forming it into a product, often thousands of times each day.

Likewise, recycling processes are also made more efficient when the incoming material is homogenous. Conversely, "contaminating materials" includes a broad group of materials – any proportion of which can derail the recycling process. Broadly defined, contaminants may include adhesives, tapes, labels, coatings, paint, overmolded or co-injected polymers, molded-in inserts, pigmented plastics, blends of polymers, or filled plastics. At high enough loadings, each additive in a plastic could even be considered a contaminant, and multiple additives in plastics should generally be avoided, especially when their functions can be integrated into the polymer backbone or the design itself.

But this guideline is not stated in a "thou shalt not" manner – it is a positive statement encouraging what *to do*. The goal essentially is to combine multiple part features into a formed part made from a single material. This means the designer should at least consider the following green design practices for meeting physical property requirements:

- *Integrate mechanical features within the plastic part itself during the part's molding operation, rather than as added components during secondary assembly.* Examples of the ways mechanical features can be integrated within a single-material design would be a hinge formed into a material as a web of flexible plastic, or as a component overmolded onto the main part in the same material. Or, using overmolding or co-injection, a separate material from a similar polymer family might be molded around or into the main molding to create mechanical effects (such as combining hard and flexible areas into one part, for instance). When this technique uses polymers that are very chemically compatible, it helps maintain the structure's recyclability.
- *Avoid added inserts, adhesives, or separate fasteners.* Metal fasteners and inserts and adhesives are contaminants to recycling, but more importantly, they themselves add environmental impacts to the part system from their own production. Molded-in snap-fits may work just as well or better than attached fasteners made from other materials. If attachment inserts cannot be avoided, ideally they should be made from plastics in the same family as the main part or they should be made from commonly recycled metals. Plastic

welding (ultrasonic, vibration, etc.) can also be used for permanent bonding without adding foreign contaminating materials to the part system.

- *Use single-layer or single-polymer layered structures, rather than multi-polymer laminated parts.* In thin packaging applications, necessary barrier layers such as ethylene vinyl alcohol are difficult to substitute for. But new polymers and additives allow for more single-layer constructions – or for multi-layer parts built from polymers in the same family. Complicating the choice of following this principle are multilayer laminated structures that combine incompatible polymers so as to use their properties in synergy. These structures may significantly reduce the total amount of plastic needed for a package. Such a structure may not be recyclable, but it still has green advantages.

For achieving desirable aesthetics using a single plastic material, these practices should be favored:

- *Favor transparent or unpigmented products, or a series of designs with a common color orientation.* Pigmented or other filled plastics are difficult to incorporate into a recycling stream. Even within a plant, scrap in multiple colors requires special handling for reintroducing it into the production process. Moreover, the pigments themselves sometimes have complex environmental impacts and handling issues in the plant. On the other hand, transparent or "natural" unpigmented plastics, sometimes made clearer with nucleating/clarifying agents, can be extremely attractive and are easier to recycle.
- *Use molded-in finishing rather than painting or chrome plating.* If a part must be colored, molded-in pigmented plastic is preferable to painting or plating. Volatile organic compounds (VOCs) and effluents from painting or plating can have very objectionable environmental impacts. Some molded-in colorants or additives even allow the part to resemble one that has been plated or specially painted [7]. Integrally colored parts do, however, face difficult aesthetic and weatherability requirements if they are to be used in automotive or construction parts exposed to sunlight – requirements that still require most exterior automotive plastic parts to be painted.

- *Consider useful kinds of molded-in aesthetics.* A variety of surface textures and grains can be molded into a part's plastic. These effects do not require a special finishing operation, but still allow a part to be differentiated or to seem unique. Part identifying codes can also be molded into the plastic itself. On the other hand, molded-in labels or decorative elements made from other materials can be difficult or impossible to remove from a part, and, depending on their composition, are likely contaminants in the part's recycling stream.
- *Choose less-contaminating secondary operations:* Some secondary processes such as laser decorating add attractive brand differentiation to a part but require no or minimal materials added to the part. And the ink printing of identifying information on plastics adds less contaminating material to a part than adhered labels.

These design choices also tend to reduce part production costs, an additional argument that justifies their sustainability.

5.1.6   Incorporate Renewable Content

Although this principle may sometimes conflict with the above principles, designers should still consider how a design might allow the use of materials that are based on renewable feedstocks. Adopting traditional resins that are bio-based should never be a difficult switch to make, but biologically-synthesized plastics have unique physical properties and molding/forming behaviors. They may require different wall thicknesses or corner radii in a part, compared to a similar part made from a traditional plastic. (This factor itself may be one of the biggest reasons processors resist trying bioresins, even though PLA grades are formulated to process in ways that are familiar to them.)

The other obvious question for a designer is whether a design of a part made from a bio-based resin would be a financially sound option to even consider, given that the bioresin likely will carry a cost premium. Such a decision would be better made with input from the marketing department and executive management than with the help of engineering.

Changing to renewable material does not have to be 100% and all-ecompassing for a product. A complete changeover is especially difficult to do for products in engineering applications. But starting with a low bio-content material is, at least, a good foundation on which to build.

Some new plastics compositions incorporate both renewable and recycled content while maintaining favorable aesthetics in a design. Efforts have been as basic as substituting natural fibers like hemp or coir for glass fiber in certain automotive applications. In addition to using renewable materials, re-using expensive material such as carbon fiber can lead to environmental and financial gains. Carbon fiber recycling not only prevents the waste of virgin carbon fiber in landfills after its first use, but components produced using the recycled fiber are themselves recyclable, because carbon can retain a significant portion of its virgin properties even after a second reclamation [8]. These incremental gains are respectable for a demanding sector such as the automotive industry.

5.2 Consumer Preferences in Green Design

Are consumers motivated to support green design goals? Might they even perhaps see themselves as participants in a company's efforts, by purchasing products that can be shown to be more environmentally sustainable? Yes and no – it is a matter of degree, at least when speaking about the performance of functional products. The more the product's cost seems excessive for its resulting performance – or the more the expected product properties seem sacrificed by green design – the smaller the proportion of consumers who are willing to collaborate in the company's efforts in sustainability.

In addition, what about the appearance of highly visible products? When should the green character of the product be made visibly obvious to the consumer, even though this goes against traditional plastics design guidelines that emphasize clear, aesthetically perfect surfaces? The part itself can become an advertisement for the sustainable efforts of its producers. As the world becomes more aware of the impact of unmanaged plastic waste on the environment, even the simplest package carries a message that it is "100% rPET" or "made from recycled bottles", or even "BPA free" when the use of BPA was never a requirement or concern. Car manufacturers are already experimenting with natural fiber containing components in which the natural fibers are visible on the surface of the part, traditionally a design "no-no." And makers of construction products have known for years that plastic-based products that mimic the textures, color tones, and grains of real wood are favored by builders and homeowners – this could perhaps be leveraged into advertising the materials bio-based content as well. A green-designed product that is fibrous or rough looking can stand out from traditional product designs, signaling "green" to the marketplace and opening up opportunities for more sustainable products and materials.

And finally, what about green design changes that a consumer cannot see or might not even be aware of? Such a product might have to include a message to the consumer explaining and even arguing for the green design change the consumer may or may not have noticed. Material changes in the design can be even harder for a consumer to notice (unless we are talking about the notoriously noisy PLA SunChips® bag discussed in Chapter 4). Such external, marketing-oriented considerations make the material-selection process, discussed in the next chapter, even more difficult than it traditionally has been for engineers and designers.

References

1. Carlin, C., AMI's Thin Wall Packaging Conference Offers a Look at Germany's Innovative Circular Economy Strategies. *Plast. Eng.*, 75, 2, 6–11 2019, February.
2. Anastas, P.T. and Zimmerman, J.B., The twelve principles of green engineering as a foundation for sustainability, in: *Sustainability Science and Engineering: Defining Principles*, M.A. Abraham (Ed.), Elsevier B.V, Amsterdam, 2006.
3. https://www.theguardian.com/environment/2019/oct/13/war-on-plastic-waste-faces-setback-as-cost-of-recycled-material-soars
4. http://petresin.org/sustainability-recyclability.asp
5. https://www.environmentalleader.com/2019/01/terracycle-circular-delivery-loop/
6. Tolinski, M., Returnables carry more weight. *Plast. Eng.*, 61, 16–18, 2005, February.
7. Tolinski, M., *Additives for Polyolefins*, Elsevier/Plastics Design Library, Oxford, 2009.
8. Surbrook, C., Recycled Carbon Fibers for Automotive Applications. *Soc. Plast. Eng.*, Recycling Division Newsletter, Q3, 7–12, 2018.

6

Sustainable Considerations in Material Selection

The process of selecting materials for a new product – or of finding alternatives to a material used in a current product – can be extremely analytical. Usually, various material physical properties for the product are identified and then weighted according to their importance in the application. Possible materials are identified and then scored in some way according to how they meet key requirements. A ratio or performance index of various key properties may be created to highlight which materials are the optimal choices (these ratios may relate strength with density, or cost per unit of some key property, and so forth). And then some final computation might be used to rank the materials according to their overall suitability in terms of properties and costs [1].

But until relatively recently, what had commonly been missing from this process were factors related to each material's environmental impact during its life cycle, from the "cradle" of the material's production to the "grave" of the product at the end of its service life. In the past decade, the

Michael Tolinski and Conor P. Carlin. *Plastics and Sustainability 2nd Edition: Grey is the New Green: Exploring the Nuances and Complexities of Modern Plastics*, (175–204) © 2021 Scrivener Publishing LLC notion of "grave" in a life cycle analysis has become somewhat polarizing since the advent of the circular economy philosophy. As mentioned elsewhere in the book, life cycle analyses are evolving to account for end-of-life scenarios – a very difficult task – because the accumulation of discarded plastics in the environment is causing harm to both natural ecosystems and marine life.

This chapter will attempt to bring into the material selection process the influence of environmental impacts of the types discussed in previous chapters for various plastic materials and applications. Mostly this chapter will make material comparisons qualitatively rather than quantitatively, given that a full-blown material selection process for even a simple product would require an entire chapter in itself, as well as very specific requirements defined for the product.

Most importantly for this book, in each section below, at least one of the material alternatives being considered is bio-based in some way, to show how these materials' environmental impacts can tip the scales in material selection or substitution.

After a brief introduction, sections that relate competing materials covered will include:

- A broad example of materials selection: plastics vs. metals and glass (6.1).
- Material selection for common high-volume applications (6.2).
  - Selection issues for PET beverage bottles
  - Selection issues for flexible and thermoformed packaging
  - Selection issues for housewares and food service tableware
- The selection of bio-based plastics (6.3).
  - PLA, PHA, TPS, and bio-based PE
  - Natural fiber reinforcement for plastics
  - Bio- and partially bio-based engineering polymers
- The selection process: A visual approach (6.4).

The decision-making process for selecting materials is never "one-size-fits-all"; each application's requirements will differ as will each company's priorities when selecting materials. Cost and property optimization are the common and expected emphases for plastics processing organizations, as are, increasingly, environmental footprint and sustainability.

Therefore, book chapters that cover materials selection should be inclusive and somewhat generalized to cover multiple situations, while also using application examples to illustrate key selection factors. Below, various application sectors are covered, and material comparisons are made mainly for the product sectors where traditional fossil-fuel-based plastics and bio-based plastics are likely to be considered as competing alternatives. Considering the growing capabilities of bio-based resins and natural fibers, this range of applications is quite wide and is constantly growing.

The tables below compare only the attributes that are most important for each material grouping or application area. Many irrelevant properties are intentionally left out; thus, for single-use packaging plastics, for example, long-term fatigue performance is not included in the comparison. Plus, to save space, the tables are further refined to show only those attributes in which there is a significant difference between the compared materials. So if all the materials being compared for an application sector can be processed with the same equipment in the same way, the category "processing cost" would not be included as a row across with which to compare them. But if a material would require a process variation or special design modifications in a part, this issue merits comment in the table and text.

The tables also include a few factors that are not normally covered during the material selection process, but which are specifically pertinent to bio-based materials. For example, as some of the tables indicate, the availability of many bio-based materials is currently limited, especially at the volumes required for the particular application covered by the table. Some natural fibers such as kenaf or coir (coconut) fiber, for instance, may have to be sourced from overseas, or their quality or availability may be affected by the specific time during the growing season when they can be harvested. Also complicating matters is the issue of constantly changing material prices for both traditional and bio-based polymers, as well as a lack of information about the full processing costs for bio-based materials.

Moreover, the use of the term bio-based in material selection must always be qualified. Some materials called "bio-based" may only be partially based on biological sources. Real bio-based content depends on what percentage of the material is based on organic carbon that comes from renewable resources, such as carbon drawn from the air via photosynthesis in recently living plants [2]. This biological carbon percentage can be determined using isotope analysis as per the ASTM D6866 standard, as discussed in Chapter 2.

Also mentioned below is the sometime controversial term *biodegradability*. Many bioplastics are inherently biodegradable, and can be shown to completely degrade during composting using test standards such as ASTM D6400. Other biodegradable plastics depend on the use of biodegradability-enhancing additives. These compositions claim to degrade in common disposal environments over a long duration. But the degree of degradation is often only estimated via extrapolations from test data (an issue also discussed more in Chapter 2). This chapter, at least, will assume that any biodegradable materials that are discussed can be confirmed to totally degrade as per standard testing.

6.1   Examples: Plastics vs. Metals and Glass

A basic way to illustrate the process of material selection is to simply compare general categories of materials such as plastics, metals, and glass. Table 6.1 shows comparisons between the general characteristics that are pertinent to an application in which these three broad groups of materials often compete (here, specifically food and beverage packaging).

Table 6.1 elucidates certain advantages plastics have over metal and glass, especially in non-engineering applications. Plastics are relatively impact-resistant, resisting permanent dents and breakage. They are inexpensive, low-density materials possessing good strength-to-weight ratios, even when compared with metals and glass. They can be attractively colored and vividly decorated without added labels. And they are processed at relatively low temperatures (although they are not easily formed or machined into complex shapes in common room-temperature processes like metal stamping).

Food-grade plastics are safe and inert for food and beverage packaging, though controversies do arise. Ironically, metal beverage containers have been associated with a plastics related health issue, since steel and aluminum cans use a liner of epoxy resin based on bisphenol-A. Here, food product makers face public scrutiny of a type that they would not face when using non-BPA-containing plastics packaging. In fact, some non-BPA containing plastic packaging is specifically and conspicuously marked "BPA free" to completely reassure the consumer [3].

Metals and glass do have their advantages, though. Compared with metals and glass, plastics have limited recyclability in current practice, with even a 50% recycling yield of a waste stream of packaging plastics being considered about as high as might be practically expected. Metals and glass can be re-melted multiple times without significant property degradation, permitting their almost complete recycling into other products (steel is the most frequently recycled of all materials). Glass and metals are also hard and stiff and serve as strong barriers to gas and moisture. They have a much

Table 6.1 General comparison of plastics, metals, and glass for food and beverage packaging.

Material Characteristic	Plastics	Metals	Glass
Physical Properties			
Density	Low	High	High
Strength per weight	High	High	Medium
Stiffness per weight	Medium	High	High
Impact resistance	Good	Fair	Poor
Barrier properties	Fair	Good	Good
Environmental/Life-cycle Impacts			
Toxicity	Medium	Medium	Low
Production energy & emissions	Low	High	Medium
Recyclability	Limited	High	High
Costs & Marketability			
Material costs	Low	Low-Medium	Low
Processing costs	Low	Medium	Medium-High
Aesthetics flexibility	High	Medium	Low
Consumer perception	Negative-Neutral	Neutral-Positive	Positive

longer history of use in consumer products than plastics, which has helped them maintain consumer confidence at a high level; accordingly, plastic products are often designed to mimic metals or glass – rarely is it the other way around.

But despite consumer confidence in metals and glass, plastics continue to replace them. There are many recent examples. New, flexible plastic pouches are designed to be reclosable and flatten out as they are emptied – advantages over stiff metal cans and their sharp edges. And lightweight, impact-resistant plastics are even now being used for wine and beer bottles and even tomato sauce, once all strongholds of glass. Along with plastic's cost and property advantages, PET bottles have also been found to be more environmentally friendly than glass bottles or aluminum cans. They use less energy to produce (creating less greenhouse gas) and generate less total waste, as shown in several studies over the past 15 years [4, 5].

Plastics are also replacing metals and glass outside of the packaging arena. A wide range of plastic types and grades offer various degrees of weight savings, reduced system costs, and more design freedom, enabled by molding and forming processes. However, the higher the engineering requirements are for an application, the less easy it is to replace metal with plastic. Often the best choice is not clear, even after one or the other material has become firmly established in an application.

Take the case of automotive fuel tanks, for example. High-density polyethylene fuel tanks have become common, replacing steel tanks starting in the 1970s. But higher requirements for blocking evaporative fuel emissions, handling new alternative fuels, and accommodating new vacuum seals in fuel tank designs, plus new steel-forming processes, are tending to make steel favored again as a tank material [6]. Other improvements in advanced high-strength steels are enabling the production of lighter weight steel structures that resist competition from plastics composites. Still, plastics loaded with higher levels of longer reinforcing fibers are becoming options for more structural, metal-like applications. Thus a material selection process comparing these materials for some applications would need to be redone every few years.

6.2   High Volume Plastics Applications

6.2.1   Beverage Bottles: PET vs. rPET vs. Bio-PET

Beverage bottle production is an area dominated by one polymer type, PET. But there is increasing competition within the application between PET materials with different life-cycle signatures – mainly different kinds of "cradles," if we are to use the "cradle-to-grave" metaphor:

- Traditional, fossil-fuel-based virgin PET offers bottle manufacturers the chance to produce bottles with the absolute lowest wall thickness for the the job, and the material is also highly recyclable.
- Bottles containing recycled PET (rPET) content are now common in many markets. However, state-of-art recycling processes are required for producing food-grade rPET with

adequate purity and proper viscosity (melt flow) for bottle manufacturing. In North America, bottle producers have been producing bottles containing rPET at 25% content or more. And "bottle-to-bottle" recycling plants likewise have started producing food-grade rPET pellets, using recycled material obtained at about a 60% yield [7].

- PET partially or wholly based on renewable, plant-based sources (bio-PET) is identical to traditional virgin PET in terms of properties and recycling. In 2009, the Coca-Cola Company expected to convert all it bottles to bio-PET "PlantBottle" materials by 2020. In fact, the company decided to allow "non-competitive" companies to use the PlantBottle technology [8]. Heinz Tomato Ketchup packaging and Ford Fusion interior fabrics are just two examples of leveraging Coca-Cola's technology. Recent news (or lack thereof) from Coca Cola about the PlantBottle suggest that difficulties remain in achieving an acceptable cost-benefit ratio [9].
- New developments in advanced, molecular recycling suggest that PET can be depolymerized into its component monomers, dimethyl terephthalate (DMT) and monoethylene glycol (MEG), and re-polymerized to create "virgin-quality" PET. This is a potentially circular approach that, if scaled, will offer processors the ability to re-use low and no-value waste PET and polyester feedstocks [10].

Any comparison between different kinds of PET needs to take into account the subtle ways in which the materials differ. The most glaring differences between the above options, of course, are in terms of their life-cycle impacts. Table 6.2 compares the three main options for PET bottles in beverage applications.

Table 6.2 would seem to indicate that plant-based PET is a sound choice – assuming the beverage producer is able to justify its extra costs. Ideally, a bottle created from a combination of bio-based PET and recycled content could be a potent mix for fulfilling both marketing and sustainability goals. Otherwise, material selection would depend on addressing the following general questions, as well as on a company's material selection goals:

1. If there is a cost premium for using food-grade rPET or bio-based PET, can a company justify this cost by marketing the bottles as "eco-friendly"?

Table 6.2 Comparison of PET materials for beverage containers.

Material Characteristic	virgin PET	bio-PET	Food-grade Recycled PET
Physical Properties			
General physical/mechanical properties	Optimum	Optimum	Tendency to vary
Melt flow/intrinsic viscosity	Consistent	Consistent	Tendency to vary
Environmental/Life-cycle Impacts			
Production energy and toxic emissions	High	High	Low-Medium
Net fossil-fuel consumption in production	High	Medium	Low
Costs & Marketability			
Material/manufacturing costs	Low	Medium-High	Low
Material availability	High	Low	Medium but growing
Consumer perception	Negative or neutral	Positive	Positive

2. What percentage of rPET in the application is feasible? At what point would a bottle design have to be changed, if at all, to accommodate a high recycled content?

3. Does rPET have adequate and consistent melt-flow/viscosity for bottle manufacturing? Would its melt flow have to be addressed using additives or some other compensation?

4. Can bio-based PET or rPET be sourced at the necessary volumes, and at affordable, stable prices?

5. Can recycling systems manage to capture and deliver enough raw material feedstock (waste) to those companies pursuing advanced recycling?

6.2.2   Thermoformed and Flexible Packaging

In other packaging applications that use thin-gauge plastics, conventional material choices may include a number of fossil-fuel-based polymers: such as PS, PP, and PET sheet for thermoformed rigid packaging, and layered constructions of polyolefins (PE and PP) for flexible packaging and blown-film bags. But increasingly, an alternative to these materials is the 100% renewable, biologically sourced material polylactic acid (PLA). Attempts to compare PLA with key competing materials will be made in this section and later in this chapter.

In thermoformed packaging, such as produce trays, clamshells, and deli trays, PP, PET and PS are common materials, though PLA has increasingly become a noticeable presence in many takeout containers and drinking cups. However, multiple factors are hindering PLA's growth, such as low impact strength (which can be addressed using additives or blended polymers, which increase material costs) and macroeconomics. Limited global supply of PLA and competing uses in woven and non-woven applications mean that demand is driving the price up. Critically, PLA does not fit the current recycling stream for reclaiming traditional plastics, and large-scale industrial composting facilities are insufficient for taking advantage of PLA's biodegradability. Meanwhile, a stream of mainly bottle-based recycled PET is available for use in improving the environmental footprint of thermoformed PET packaging. Table 6.3 makes some basic, relative comparisons between the three materials, reflecting an uphill climb for PLA unless its green qualities are prioritized and priced into more applications successfully.

Flexible packaging films have some similarities with thermoformed packaging. Here, in applications such as sealed food bags and shrink

Table 6.3 Comparison of PE, PP, PET, PS, and PLA for thermoformed or flexible packaging.

Material Characteristic	PET	PE & PP	PS	PLA
Physical Properties				
Density	High	Low	Medium	High
Impact/crack/tear resistance	Good	Good	Poor-Fair	Poor without additives
Heat resistance	Poor-fair	Fair	Fair	Poor
Processing consistency & confidence	High	High	High	Low
Environmental/Life-cycle Impacts				
Production energy and toxic emissions	Medium	Low	Medium	Low
Bio-based/renewable content	Low	Low	None	100%
Recyclability*	Medium–High	Medium	Low-Medium	Low
Biodegradability/compostability	None	None	None	High
Costs & Marketability				
Material/manufacturing costs	Low	Low	Low	Medium–High
Material availability	High	High	High	Limited, but growing
Consumer perception	Negative to Neutral	Negative to Neutral	Negative	Positive

*"Recyclability" means not only the technical ability to recycle a material, but the availability of systems to collect, sort, and reprocess it.

sleeves, PLA has some potential as an alternative choice. But another of its Achilles' heels is its poor moisture- and gas-barrier properties, remedied only by a significant barrier-film layer or additive materials. Metallized PLA has been used in snack food bags; these products try to capitalize on PLA's environmentally friendly bona fides. Yet layered films of poly-olefins (PE and PP) and other commodity plastics remain obvious low-cost choices for flexible packaging and film. Polyolefins have relatively low envi-ronmental impacts, as shown in Table 6.3, and greater amounts of poly-olefins produced from sugarcane ethanol will create alternative bio-based materials with an even lower environmental impact, though perhaps in limited, but growing, supply.

Given all the variations of flexible and thermoformed packaging that exist, more specific requirements would have to be added to Table 6.3 to make a full list of practical material selection criteria. Issues about gas- and water-barrier properties, clarity, and heat resistance may influence mate-rial selection even more than most of the above general considerations, depending on the application. But the comparison here does indicate some general questions that should be answered even before the material selec-tion process starts:

1. Does the retailer of the packaged product (e.g., Walmart) have special renewable-content goals for its packaged prod-ucts that could be met with the choice of bioresin packaging?
2. Does the packaged product itself, such as organic food, have green associations for consumers that would make a bioresin an attractive, marketable choice?
3. In a material substitution situation, can the bio-based mate-rial be processed on the same equipment as traditional resins? (This is also helpful to know if the sourcing of the bioresin at required volumes is uncertain or proves to be inconsistent.)

Example cases: The growth of PLA would appear to speak to its suc-cess in the market, competing with other grades of plastic include rPET. GreenWare™, produced by Fabri-Kal, is one notable example and is found in many retail locations in the US. Continued research and development in the coffee pod sector has led to breakthroughs in temperature stabil-ity for crystallized PLA. Yet several brands, notably Tesco in the UK, and converters have declined to pursue the use of PLA in packaging for life-cycle reasons, primarily due to the difficulties in managing end-of-life concerns [11].

6.2.3   Housewares and Food Service Tableware

Broadly defined, "housewares" covers multiple items and tools in every-day use. They are often expected to be durable enough to last for years of use, and when made of plastic, are expected to be inexpensive. Plastic housewares include kitchen tools and utensils, washable storage containers and cups, bathroom accessories, toys, hangers, and hooks and hardware for light use. Given their durability requirements and their contact with water and food, plastic housewares are expected to be made from traditional, inert, nondegradable materials such as polypropylene, polycarbonate, ABS, and thermosetting polymers, when heat resistance is required.

Until more durable bioplastics become widely available to handle these applications, housewares manufacturers will continue to rely on traditional materials. But "green plastic" opportunities exist in housewares. The bacteria-produced bioresins in the polyhydroxyalkanoate (PHA) family, having polypropylene-type properties, are making progress in this sector [12]. Another way of greening housewares is to produce them with more recycled content, a trend also starting to gain force.

However, this section will mainly focus on the selection of materials for housewares applications in which bioresins are used more widely: single-use food service tableware such as plates, cups, and utensils that are intended to be disposed of after use. Plastics such as polystyrene and polyolefins have been used for years for these purposes; now, bioresins are seen as marketable options. Biodegradable bioresin tableware is attractive to consumers who feel satisfied knowing that the cup or plate they just discarded was designed to eventually degrade and return to nature, and that fossil fuel resources were not wasted to create the disposable item. Plant-based resins such as thermoplastic starch (TPS) and PLA and other plant-based plastics have adequate properties for use as knives, forks, spoons or plates. Ideally, biodegradable bioresin tableware is particularly effective when used in high-volume food service environments (such as cafeterias), where used plates, cups, and utensils can be collected and sent with food scraps to an industrial composting facility.

The selection table for food service tableware materials (Table 6.4) is much like the previous table for thin packaging, except that PET is excluded, and multiple biodegradable bioresins (PLA, PHA, and TPS) are included, though evaluated as a unit. In this application group, as in others, a positive consumer perception would require the consumer to somehow be informed about the biocontent of the items they are using, which this comparison assumes.

Table 6.4 Comparison of PE, PP, PS, and Biosresins (PLA, PHA, TPS) for disposable food service tableware.

Material Characteristic	PE & PP	PS	Bioresins
Physical Properties			
Density	Low	Medium	Medium-High
Heat resistance	Fair	Fair	Adequate
Processing consistency & confidence	High	High	Low
Environmental/Life-cycle Impacts			
Production energy and toxic emissions	Low	Medium	Low
Bio-based/renewable content	Low	None	Up to 100%
Recyclability	Medium	Low	Low
Biodegradability/ compostability	None	None	High
Costs & Marketability			
Material/manufacturing costs	Low	Low	Low-High
Material availability	High	High	Limited, but growing
Consumer perception	Negative	Negative	Positive

In the selection of bioresins for tableware, the greatest environmental gains will be made when large-scale food service operations and their suppliers make a complete shift from traditional resin to bioresin products. But the following questions would need to be answered by one or both of these groups before making such a transformation:

1. What must the minimal environmental gains be to make a shift justifiable? (Is touting the renewable origins of the

products enough, or would their use need to be linked to *real* industrial composting, for instance?)

2. What about bioresin products that are only partially bio-based? What about non-biodegradable bio-based resins? Can these be honestly and effectively marketed as having green benefits?

3. How can consumers be made fully aware that they are using bio-based tableware products? Are the benefits of making a switch to bioresins justified even if consumers are not fully informed that it has happened?

Example case: Bioresins such as PHA are now used in starch-based bioresin blends for compostable tableware such as forks, knives, spoons, plates, and cups. The materials are designed to degrade within 180 days or less in composting, or within 2–3 years in a landfill [13]. This satisfies the minimal environmental demands questioned in the first question above, except that industrial composting options are limited for handling used tableware. The third question above can be satisfied by clear labeling on the product packaging, symbols on the products themselves, or signage where the tableware is provided. Otherwise it might be difficult to justify the extra costs of this application, given that these products appear to be very similar to other disposable tableware, at least in the eyes of the average consumer.

6.3   Bio-Based Plastic Selection

Material selection is being made more and more difficult by the growing numbers of bioplastics coming into the market, and by their unfamiliar properties. The following subsections group traditional and bio-based plastics with similar properties, applications, and polymer families, aiding in making comparisons and for selecting between them.

6.3.1   Bio-Based Resins: PLA, PHA, TPS, PE

This section will offer some background for comparing three modern biologically synthesized resin families: PLA, TPS, and polymers and copolymers from the PHA family of bacteria-synthesized resins. Table 6.5 also includes representative characteristics of bio-based polyethylene, which is now being produced from sugarcane ethanol and has properties that are equivalent to fossil-fuel-based PE.

Complicating matters for bio-synthesized resins are the additives and blends that are needed for allowing TPS, PLA, and PHA to compete better with traditional commodity resins. Additives can improve impact strength, melt strength, thermal stability, crystallization nucleation, and other key properties. Bioresin additives preferably should not reduce a bioresin's percentage of bio-based content, its biodegradability, or especially in the case of inherently transparent PLA, its clarity. Moreover, when additive masterbatches are used to mix additives in with the biopolymer, biodegradable carrier resins should be used. Companies such as Clariant (Muttenz, Switzerland) and Novamont (Novara, Italy) have successfully commercialized both conventional and bio-based color masterbatches with colors derived from natural sources, e.g. orange curcuma from turmeric, yellow urucum from tropical flowers, and green from chlorophyll in many plants [14].

So for simplification, the biopolymer properties compared in Table 6.5 are based on materials that are free of complex additives. But commercial reality requires more complex thinking; thus it should always be remembered that the three key modern bioresins here should still be thought of as *resin systems* of mixed materials, rather than as individual polymers:

- *TPS:* Along with tableware, TPS has been used for bags, packaging film, display trays, and mulch film. Starch-based material has acceptable melt strength for processing, but it can be brittle, especially in dry environments. At the same time, TPS absorbs moisture quickly and then degrades readily, limiting it to short-duration, superficial applications. So TPS must be modified to improve its heat resistance and allow its use in conventional plastics processing and forms. To expand its range of use, TPS can also be blended with LLDPE for film, or with high-impact PS for injection molding. Rigidity can be added to TPS via blending with the biodegradable bioresin polycaprolactone or ethylene vinyl alcohol [15].

- *PLA:* Despite its growing range of uses, PLA has limiting factors that are only addressed with additives. PLA is sensitive to thermal and hydrolytic degradation in processing, causing degradation. As mentioned above, PLA's gas- and water-barrier properties are inadequate for many packaging uses without barrier layers or other added materials. And pure PLA lacks the melt strength required for blown-film

Table 6.5 Comparison of biosynthesized polymers with sugarcane-based polyethylene.

Material Characteristic	TPS	PLA	PHAs	bio-PE
Physical Properties				
Density	High	High	High	Low
Strength & stiffness properties	Low	Medium	Low-High	Low-Medium
Heat resistance	Poor	Poor	Fair	Fair
Processing consistency & confidence	Fair	Improving	TBD	High
Environmental/Life-cycle Impacts				
Transportation energy from production/ cultivation region*	Low	Low	Low	High
Recyclability	Poor	Limited	Possible	High
Biodegradability/compostability	High	Medium	High	None
Costs & Marketability				
Material/manufacturing costs	Low	Medium to Low	High	Medium to Low
Material availability	Abundant	Growing	Limited	Limited
Consumer perception	Positive	Positive	Positive	Neutral-Positive

*Assuming production of the material in the southern hemisphere and conversion and use of the plastic in the northern hemisphere.

extrusion and foaming, plus it has low impact strength, brittleness, poor tear strength, and low heat deflection. PLA's brittleness can be reduced with bio-based plasticizers such as sorbitol or glycerol, though many impact modifiers do not meet American Society for Testing and Materials (ASTM) compostability guidelines above certain use levels. Its toughness can be improved using engineered mineral fillers, and nanoparticles have been used as an impact-strength modifier, reportedly without affecting PLA's clarity in packaging applications. Sukano, a Switzerland-based developer of masterbatch additives, has created impact modifiers, slip and antiblock agents for PLA that increase usability without affecting temperature resistance [16].

- **PHA:** As one useful form of bacterial PHA, PHB (polyhydroxybutyrate) and its copolymers have potentially a wider range of uses than PLA, with properties that approach those of a semi-engineering polymer like polypropylene. And PHB is more biodegradable than PLA [11]. Moreover, useful PHB grades may prove to be less dependent on additives than PLA, since microbial engineering can be used to produce various bio-copolymers, such as the flexible PHBV (poly[hydroxybutyrate-*co*-hydroxyvalerate]). PHAs can also be blended with other less expensive biomaterials to create tough, useful, biodegradable materials. starch, and soy protein to produce material for rigid cups.

Also included in Table 6.5 is a natural competitor for bio-synthesized resins: bio-based polyethylene based on monomers derived from sugarcane ethanol. HDPE and LDPE produced in this way are technically 100% bio-based (depending on their additives), with properties equivalent to any fossil-fuel-based PE. Bio-PE also processes and is recyclable like traditional PE, but unlike PLA, TPS, and PHAs, it is not at all biodegradable. Its availability is limited by the amount of land that can be devoted to sugarcane grown for creating industrial ethanol. However, PE's long history of use, consistent properties, and high confidence in processing will likely, in coming years, make bio-PE a sought-after material selected by traditional processors.

Along with issues discussed above, the complexities of bioresin selection factors create the following additional stumbling blocks for a material selector to address:

1. How can a bioresin-containing plastic be labeled or evaluated in terms of sustainability if it is not 100% bio-based?
2. How important is biodegradability for a material, given that industrial composting facilities are rare?
3. And when is a 100% bio-based, recyclable, non-biodegradable polymer (such as sugarcane-based PE) more justifiable for an application, compared with a renewably-sourced biodegradable polymer?

Example cases: Because bioresins have limitations in providing many of the properties provided by traditional resins, materials selectors will be tempted to consider using blends of bio- and fossil-fuel-resins (this issue relates to the concern of question 1 above). Some compromise may be needed when evaluating these resin blends' green attributes against needed properties. Meanwhile, more of these blends are becoming available. Consider the following examples:

- Over the past decade, BASF has developed Ecoflex® and Ecovio®, two bioresins with distinct characteristics. The former has been on the market for over 20 years, is fossil-based, is certified as biodegradable and compostable, and is used in many bioplastic blends. The latter is certified as compostable with variable biobased content. Primary uses included organic waste bags and agricultural films, though it is also finding favor in the development of compostable coffee capsules [17].
- Sukano (Schindellegi, Switzerland) has developed additives for PLA including a transparent modifier S633 as used in thermoformed packaging. This nucleating masterbatch forms many small nucleation sites which increases the speed of crystallization. A melt strength enhancer improves IV and mechanical properties while retaining transparency. Using LDR of 1-3% is shown to reduce brittleness without affecting temperature resistance. For injection molding grades of PLA, Sukano S687-D, an opaque impact modifier, improves elasticity and toughness for high-stress applications when used at 10-30% dosage rates [16].

Questions remain about whether these examples can ethically be marketed as a bio-based materials – even if the material grade includes 51% bio-based resin, or 60%, or 80%. Programs such as USDA Biopreferred and similar guidance in the EU have been developed to provide standards and regulations to help manufacturers understand how to create bio-based materials that are eligible for government procurement programs [18]. But apart from following government definitions or standards, what percentage of biocontent would interest what percentage of consumers (not to mention the question about whether significant sustainability goals are actually being achieved with the blended materials)? And could products made with the bioresin be successfully marketed with a message for consumers about its bio-based percentage, even if it is low?

6.3.2   Natural Fiber Plastics Reinforcement

The property limitations of both biopolymers and traditional polymers can be extended using reinforcing fibers. When the fibers are made from natural materials, they add bio-based percentage content to the plastic. This subsection will attempt to evaluate various natural fibers that might be selected, at times comparing them against the baseline non-biological fiber in common use: glass.

Natural fibers, mixed with traditional polymers, at least allow some opportunities to create parts that are partially bio-based, according to the percentage of natural fiber in them. Moreover, natural fibers incorporated in bio-based polymers offer the ultimate goal of structural composites that are 100% bio-based as well.

Natural fibers, or "bio-fibers," include mainly plant-based fibers such as flax, hemp, wheat straw, kenaf, jute, coir, abaca, wood (in various forms), and the other fibers covered in Chapter 3. Developmental work using each fiber in various plastics has led to applications of natural fiber composites (NFCs, sometimes also called "biocomposites" even if the matrix polymer is not bio-based). Alongside the common wood-plastic composites used in building and construction are various NFC automotive parts, including door panels and other interior components. Unlike the wood-filled plastics used in bulky construction materials (where weight is not a big issue), NFCs can also be formed into thin, lightweight, contoured parts with various degrees of strength and rigidity.

NFCs contain fibers in various forms, in woven or nonwoven mats, as twines, or as long- or short-chopped fibers. Fibers may be loaded into NFCs at percentages of up to 70%, sometimes incorporated in NFC sandwich structures composed of various layers of fiber mats and polymer. These layers are then compression-molded together to create lightweight panels. NFCs can also be created with other low-pressure molding and extrusion processes.

Because natural fibers are sensitive to high-temperature/high-pressure processing, NFC matrix materials are usually limited to lower temperature thermoplastics (such as polypropylene) and some thermosets. In a way, natural fibers' temperature and forming pressure limitations can be viewed as advantages in processing, requiring less energy and lower-strength tooling for molding or forming NFC parts. But natural fibers do tend to absorb water, meaning that processing must control moisture content at the proper low levels. By contrast, glass fibers do not have these temperature and moisture issues, though they are difficult to handle and abrasive to both tooling surfaces and skin.

Also in contrast to glass, natural fibers generally have lower costs, lower production energy requirements (being extracted directly from raw plant material), and lower density. In compensation, they do not provide as great an increase in mechanical properties as glass fibers do when used to reinforce polymers. But reinforcing fibers of all kinds can increase a plastic composition's tensile strength, stiffness, and impact resistance.

Given that they are natural materials, natural fibers are less consistent in dimensions and overall quality compared with industrial glass fibers. Some natural fibers, such as abaca (banana) fiber and coir (coconut) fiber are simply plentiful waste materials from agricultural processing. Here, the main production issues are separating pure fibers from the waste and transporting the fiber from sometimes distant locations. Natural fibers are, of course, biodegradable, though only when exposed to the environment or used within a biodegradable polymer. NFCs offer limited potential in recycling. Table 6.6 shows some of the basic comparisons between a few natural fibers and glass fiber, showing a wide range of properties.

Despite the range of properties of natural fibers indicated by the table, property distinctions between NFCs based on different fibers in the same polymer matrix can actually be quite subtle. The number of natural materials that are being used in plastics is ever-expanding, so when considering natural fibers and fillers, material selectors should always check on current developments. Materials once considered nuisance industrial wastes are being found to have useful properties as fillers or reinforcements. Principles of green chemistry and industry ecology dictate that these waste streams should be turned into useful products when possible. Examples from over a decade ago are still relevant and have served as guidance and inspiration for new experiments. Waste materials such as ground soy stems and hulls [19] and even ground waste leather from shoe production [20] can be used as fillers in plastic and rubber, respectively. Using cellulose for mechanical reinforcement has been shown to increase the strength and Young's modulus or all polymer matrices [21]. Such "outside the box" ideas

Table 6.6 Comparison of glass and natural fibers for reinforcing plastics [22].

Fiber Characteristic	Glass	Flax	Sisal	Coir (coconut)
Physical Properties				
Density	High (2.5)	Low (1.4)	Low (1.3)	Low (1.3)
Tensile strength (MPa)	High (2400)	Medium (800–1500)	Medium (600–700)	Low (200)
Tensile modulus (GPa)	High (73)	High (60–80)	Medium (40)	Low (6)
Elongation at break	Medium (3%)	Low (1.2–1.6%)	Medium (2–3%)	High (15–25%)
Heat resistance	High	Low	Low	Low
Processing consistency & confidence	High	Medium	Low-Medium	Low
Environmental/Life-cycle Impacts				
Production energy	High	Low	Low	Low
Transportation energy from cultivation region*	n/a	Medium	High	High
Recyclability	Medium	Low	Low	Low
Biodegradability/ compostability**	None	High	High	High
Costs & Marketability				
Material/ manufacturing costs	Medium	Low/Variable	Low/Variable	Low/Variable
Material availability factors	None	Seasonal	Seasonal	Seasonal
Consumer perception	Negative/Neutral	Positive	Positive	Positive

*Assuming NFC manufacturing takes place in the temperate northern hemisphere.
**Assuming fiber is exposed to environment or is incorporated in a biodegradable polymer matrix.

on how to incorporate natural waste materials with biological polymers is another step toward achieving durable products that manufacturers can claim use renewable resources extremely efficiently.

6.3.3 Engineering (Bio)polymers

Durable products require engineering polymers with higher strength, stiffness, and thermal resistance. Natural (or even glass) fiber reinforcement in commodity materials like PP normally cannot reach the properties required for engineering applications. Unfortunately, the options are limited for material selectors seeking bio-based engineering plastics. Truly bio-derived engineering polymers are few, at least partially because the relatively low volumes in which engineering polymers are used do not motivate efforts in developing engineering biopolymers.

There is another factor that de-motivates their development: Consumers do not think of engineering plastics as being truly wasteful materials, so there is low interest across the board in bio-based substitutes. After all, unlike packaging polymers, engineering polymers are used for long-term components hidden from view inside complex devices; thus they are less often directly handled and disposed of by consumers. By comparison, the environmental impact of packaging plastics can be seen everywhere simply by looking into any public waste receptacle.

Still, interest (and investment) is growing in bio-based materials of all kinds, and polymer suppliers have gradually been developing engineering resins that are at least partially based on renewable resources. Some of these grades are "hybrid" blends of PLA and traditional engineering polymers such as polycarbonate and polymethyl methacrylate; others, including polyamides (nylons) and polyesters, are at least partly based on feedstocks derived from renewable sources. Although the properties of these new options may not yet reach the level of traditional engineering polymers, a goal of these suppliers is to create a selection of bio-based materials that could serve as "drop-ins" for traditional polymer applications. Some hybrid materials and related engineering materials are shown in Table 6.7, with key properties and their suppliers' estimate of their percentage of bio-based-content.

Certain kinds of nylon, one of the most common families of engineering polymers, are based on renewable resources. As discussed in Chapter 3, nylon 11 is 100% based on oil from castor beans, and nylon 6/10 is partly based on castor oil, giving it a biocontent of about 60% by weight. Until recently, both resins were limited only to specialty applications, rather than as replacements for the common nylons 6 and 6/6. Nylon 11 is useful

Table 6.7 Comparisons of engineering resin grades containing bio-based content [23–25].

Material (and Specific Grade)	Percent Bio-Based Content	Density (or Specific Gravity)	Tensile Strength (MPa)	Tensile Modulus (GPa)	Impact Strength (Notched Izod, 3.2 mm) (J/m)	Deflection Temperature at 66 psi (°C)
PLA/PC (RTP 2099 X 121235D)	32%	1.18	48	2.1	881	121
PLA/ABS (RTP 2099 X 121236A)	40%	1.12	54	2.6	27	71
PLA/PMMA (RTP 2099X 115375C)	40%	1.21	69	3.4	37	66
Nylon 11 (Arkema Rilsan® PA 11)	100%	1.02–1.03	37–58	1.1	7–15**	n/a
Nylon 6/10 (DuPont Zytel® RS LC3060 NC010)	60%	1.07	58*	2.0	6.8**	n/a

*Yield stress.
**Charpy notched impact at room temperature; kJ/m².

primarily for highly-demanding fluid-handling components, for example, but now has been formulated for 3D printing applications, high-performance sports shoes, and light-weighting of automotive engine components [26]. Partially castor bean-based content is used for recent grades of DuPont™ Zytel® RS nylons for automotive radiator end tanks [27]. Properties for Arkema's Rilsan® nylon 11 and DuPont 6/10 grades are given in Table 6.7.

Material selection for engineering applications obviously depends more on meeting rigorous property requirements of a certain application than on life-cycle impacts and consumer perceptions. But manufacturers of materials and engineered systems are capitalizing on advertising the sustainable content of their plastics when an appropriate bio-based material does fit its property requirements, at acceptable costs. The value of using a material with a lower environmental impact has yet to be quantified as a reliable numerical value that can be incorporated into material selection, and probably such a metric cannot be created. But this should not stop material selectors from weighing life-cycle impacts as part of the selection process.

6.4   The Selection Process: A Visual Approach

Given that quantitative algorithms may be difficult to create for comparing plastics in terms of sustainability, material selectors lack a "true" scoring system that evaluates materials in terms of environmental footprint and other characteristics. Such quantitative methods would probably even be misleading. They may project false authority by calculating numerical values, with a high degree of uncertainty, which are supposed to indicate the best material choices. A more honest, and perhaps just as useful approach, is a qualitative or visualized comparison which connects a company's goals with the characteristics of competing materials. Overall, the comparisons above and the previous chapters' discussions support the construction of a hierarchy of environmentally friendly plastics, like the idealized inverted pyramid graphic proposed in Chapter 2 (Figure 2.1). In Figure 6.1 below, we propose a more detailed, slightly more "real-world" version of Figure 2.1, showing lower-impact/higher-social acceptance polymers in the top section, whose area indicates the large volumes and effects of their use. Here we assume that polymer processes will continue to become more eco-efficient and innovative, allowing more biofeedstock-based processes to produce more fully bio-based polymers that are indistinguishable from their fossil-fuel-based competitors.

Obviously, many factors determine material selection for any situation, and a material cannot be chosen simply by using Figure 6.1. Material

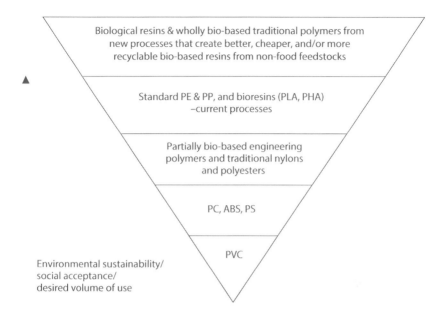

Figure 6.1 Revised "inverted pyramid" hierarchy reflecting the relative environmental impacts, social acceptance, and practical importance of commercial polymers – currently and in the future.

selection is application-dependent; thus some may argue that it is inappropriate to place PVC at the bottom of the chart, given its effectiveness in building and construction applications, for example. Nonetheless, signs, trends, and data support this hierarchy as being a reasonable guide for meeting sustainability goals being pursued by companies that rely on plastics.

A different kind of triangular, semi-quantitative visualization might also be useful when making comparisons between similar plastics for an application. In manufacturing and design, material comparisons are often made by relating their basic attributes of price, performance, and processing. These three factors can be integrated in a visual such as Figure 6.2. Similar plastics can be plotted on the chart according to how well they measure up against each other and the "three P's" (processing, performance, and price). But because our interest is in environmental sustainability, the chart should incorporate life-cycle and cradle-to-grave factors, as well as standard material characteristics. Thus, "Processing" should factor in the nature of the polymer's feedstock, and "Performance" should take into account the material's impacts during its use-life *and* during its disposal/recycling phase. "Price" in this chart remains a simple factor – the cost of the resin required in a product, with lower-cost resin options plotted closer to the "Price" vertex.

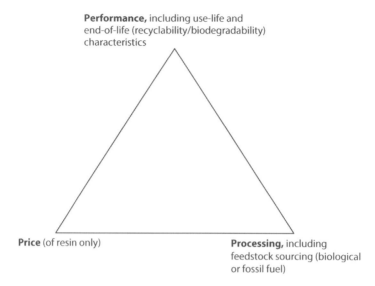

Figure 6.2 Diagram for mapping and comparing plastics materials in terms of price, processing, and performance, incorporating environmental impact characteristics.

In the use of Figure 6.2, a baseline material would first be plotted inside the triangle at a position that indicates its various attributes. For example, a simple commodity plastic's value is generally its low price, and thus it would be plotted relatively close to the price vertex. A more heavily engineered commodity plastic option would be plotted farther away from price and closer to the performance vertex.

Figures 6.3 and 6.4 are examples of two fundamental comparisons mentioned in this chapter. (Note that these charts should only be used to compare relative qualities of similar materials used for the same purposes, not absolute differences between completely different materials.) The simpler chart in Figure 6.3 compares low-priced, traditional PE with bio-based PE. The two materials are equivalent in performance properties (and recyclability). The key differences are in price (bio-PE's is higher) and overall processing profile (enhanced here by the bio-basis of bio-PE, even though it processes the same way as PE in the conversion process). Thus the bio-PE point is plotted farther from the price vertex and closer to the processing vertex.

In the example in Figure 6.4, PET, bio-based PET, and PLA are compared for use in the same application, such as an injection/blow-molded water bottle. This example assumes that all three materials have adequate

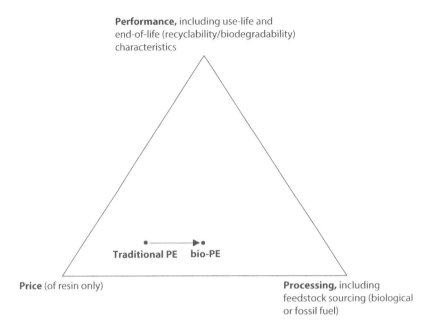

Figure 6.3 Diagram for mapping and comparing PE and bio-PE for the same application.

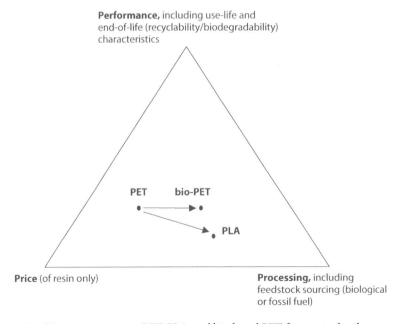

Figure 6.4 Diagram comparing PET, PLA, and bio-based PET for a water bottle application.

material properties for the application and can be processed in conventional processing. Here the PET/bio-PET comparison is similar to the PE/bio-PE comparison. With PLA, the chart accounts for its higher resin price and lower end-of-life performance overall (although PLA is biodegradable and recyclable, few industrial facilities exist for composting PLA waste or for recycling it; for these reasons alone it is plotted farther from the performance vertex). However, PLA's processing profile is enhanced by its biological basis. In the future, one would assume improved production efficiency will make PLA or bio-PET a more attractive candidate price-wise, shifting its plotted point towards price.

The placement of the materials on this graphic can be drastically altered when considering just a slightly different application. For example, for a beverage or condiment bottle that requires significant gas-barrier properties, a PLA bottle may not even be a possible candidate because of its barrier properties. Or if it could be considered, it would require significant barrier layers or added materials, relative to what is required by PET. So if PLA were considered in such a case, its negative price or performance characteristics could potentially push it far from PET's strong price placement on the chart, or perhaps totally off the chart as an inappropriate material option to consider.

Such comparisons, being only somewhat quantitative, still give a vision of what options are available for an application and how they might be compared. A low-cost material would not be expected to have a strong performance or environmental profile, but perhaps it still fulfills a company's limited goals. Or, a material selection team looking for more sustainable options may limit its view only to materials plotted closer to the performance and processing vertices. No matter what selection system is used, the decision must be recognized as having inherent complexity, so that quick, knee-jerk decisions are avoided.

References

1. Dieter, G.E., ASM Handbook, in: *Materials Selection and Design*, vol. XX, CRC Press, Boca Raton, FL, 1997.
2. Tolinski, M., Testing bioplastics. *Plast. Eng.*, 63, 6, 44–46, 2007, June.
3. https://www.nestle-watersna.com/en/who-we-are/frequently-asked-questions/are-your-plastic-bottles-bpa-free
4. Lifecycle Impacts of Plastic Packaging Compared to Substitutes in the US and Canada, in: *Theoretical Substitution Analysis*, Franklin Associates, April 2018.

5. Plastics & Sustainability, in: *A Valuation of Environmental Benefits, Costs, and Opportunities for Continuous Improvement*, Trucost, July 2016.

6. Sheffield, R. and Mould, P., Benefits of steel fuel tanks for gasoline-powered and hybrid vehicles (presentation slides), in: *Great Designs in Steel (conference proceedings)*, American Iron and Steel Institute, 2008.

7. https://www.plasticsnews.com/article/20180524/NEWS/180529946/carbonlite-eyeing-allentown-for-east-coast-plant

8. https://www.coca-cola.eu/news/sharing-plantbottle-technology-with-the-world/

9. Plastic Surgery, in: *Plastics in Packaging*, p. 7, Sayers Publishing Group, September 2020.

10. Loop Industries, company presentations, website.

11. Carlin, C., AMI's Thin Wall Packaging Conference Offers a Look at Germany's Innovative Circular Economy Strategies. *Plast. Eng.*, 75, 2, 6–11 2019, February.

12. https://www.plasticstoday.com/packaging/what-s-new-pha-bioplastics-update-cambridge-consultants/188162939961111

13. https://www.plasticstoday.com/packaging/pha-bioplastics-tunable-solution-convenience-food-packaging/157388153458558

14. Sueltemeyer, J., Colour and Additive Solutions for Bioplastics. *bioplastics Mag.*, 03/2020, 14–15, March 2020.

15. Khan, B., Niazi, M.B.K., Samin, G., Jahan, Z., Thermoplastic Starch, in: *A Possible Biodegradable Food Packaging Material – A Review*, Wiley Online Library, July 2016.

16. Carlin, C., Innovation Takes Root in California. *Plast. Eng.*, 74, 10, 8–13 December 2018.

17. www.plastics-rubber.basf.com/global/en/performance_polymers/products/ecovio.html

18. www.biopreferred.gov; https://ec.europa.eu

19. Ng, Z.S., Simon, L.C., Erickson, L., Marquis, L.A., Physical properties of recycled polypropylene with agricultural fillers, GPEC 2009 (Proceedings). *Soc. Plast. Eng.*, 2009.

20. Ferreira, M.J., Almeida, M.F., Freitas, F., New leather- and rubber waste composites for use in footwear (10.1002/spepro.002929). *Soc. Plast. Eng. Plast. Res. Online*, http://www.4spepro.org, 2011, 2010.

21. Diallo, A.K., Lentzakis, H., Drolet, R., Tolnai, B., *Mechanical Reinforcement with Cellulose Filaments*, SPE Online Technical Library, May 2018.

22. Data adapted partly from: Wong, S. and Shanks, R., Biocomposites of natural fibers and poly(3-hydroxybutyrate) and copolymers: Improved mechanical properties through compatibilization at the interface, in: *Biodegradable Polymer Blends and Composites from Renewable Resources*, L. Yu (Ed.), John Wiley & Sons, Hoboken, NJ, 2009.

23. RTP Company, Bioplastics added to RTP Company's specialty compound product families, (Press release), http://www.rtpcompany. com/news/press/ bioplastics.htm, 2011, n.d.
24. DuPont, Zytel® RS LC3060 NC010, (Product data sheet), http:// www2.dupont. com/Plastics/en_US/assets/downloads/product/ zytelrs/ZYTEL_RS_LC3060_ NC010.pdf, 2011, 2008.
25. Arkema, Key Physical Properties of Rilsan® Polyamides (Metric Units), http:// www.arkema-inc.com/index.cfm?pag=1036, 2011, n.d.
26. Hanrahan, K., Written correspondence between editor and Arkema employees, June 2020.
27. Miel, R., Castor-oil to help cut CO_2 emissions in Denso's Zytel radiator end tanks, in: *Plastics News*, p. 4, 2009, January 19.

7

Processing: Increasing Efficiency in the Use of Energy and Materials

Producing plastic products using the latest low-footprint raw materials has less importance if the manufacturer's machines and processes are themselves inefficient. Mechanical and heat energy for melting and forming resin, plus water for cooling, are critical parts of plastics processing. Optimizing their use serves the bottom line and reduces a plant's overall environmental footprint. Likewise, efficient plant practices in using reclaimed scrap and recycled material not only can save money, but also can have the same impact as purchasing bio- or recycled-content raw materials from outside sources.

Thus, this chapter will discuss issues of process optimization and state-of-the-art technologies for addressing them. It takes a long view and includes ideas that may seem "outside of the box" and unconventional for

Michael Tolinski and Conor P. Carlin. Plastics and Sustainability 2nd Edition: Grey is the New Green: Exploring the Nuances and Complexities of Modern Plastics, (205–224) © 2021 Scrivener Publishing LLC

traditional companies, especially ideas that make the longest-term impacts on a facility's environmental footprint; its sections include:

- Optimizing the recycling of in-house scrap, post-industrial material, and post-consumer plastic (7.1).
- Optimizing plastics processes for sustainability in terms of water and energy use (7.2).
  - Refurbishing used equipment and buying new sustainable machinery
  - Producing "green energy" for plastics processing

This chapter will attempt to address these difficult questions by presenting an overview of the more recent practices and technologies for converting raw resin into products more efficiently, and for cutting energy and water usage. The chapter will start by covering some of the newest technologies in plastics recycling for readers who want to understand the challenges being addressed in this area. In terms of process technologies, the main areas of thermoplastics processing equipment will be covered, illustrating general principles that are adaptable to all forms of plastics conversion processes.

7.1　Optimizing Resin Recycling

7.1.1　Reprocessing Scrap and Post-Industrial Material

Because resin costs are 70% or more of the total cost to produce a typical plastic product, the reclamation and recycling of process scrap material is a fundamental, economical practice in most plastic molding, extrusion, or thermoforming operations. It is also a very commonsensical green practice.

Obviously, the best way to handle process scrap is to not produce it. This requires focusing employees' attention on the issue through the use of environmental policies and revised procedures for reducing waste. Many companies have adopted *kaizen* practices to identify and reduce sources of energy waste, just as car manufacturers did decades ago when improving assembly line operations.

However, in plastics conversion processes, some scrap is inevitable. In-plant reclaimed material may include out-of-spec scrapped parts, short shots, sprues and runners, edge trim (or other trimmed-off material), and process start-up/shut-down material. Reclamation requires shredders or granulators that reduce these bulky materials into regrind – particles that are small enough to be fed back into melt-processing equipment.

Traditional rules of thumb say that thermoplastic products should be able to accommodate about 10–30% regrind content without affecting product properties or appearance significantly (assuming that the regrind/scrap material is kept clean and segregated from other materials, and was not scorched by excessive heat in its previous processing). Some processes, such as extrusion for sheet thermoforming, may require sheet to contain up to 50% regrind, to accommodate all the trim scrap generated by the process. This is well-understood in extrusion/thermoforming, has the benefit of allowing the processor to promote her products as more environmentally-friendly than 100% virgin parts.

In some parts, processes, or materials, the effect of adding regrind to virgin resin is inconsequential, especially if plant personnel use a consistent procedure for handling it. But blending regrind in with virgin resin can present challenges in processing, such as in the following cases:

- The regrind material may be in a form that is difficult to feed smoothly into the process, especially if it is from a low-density bulk form such as film, foam, or fiber (given their high surface areas, these forms are also prone to contamination). In-line or off-line repelletizing or scrap-densification systems can be used to convert low-density scrap into pellet-like forms, while other systems can feed trim scrap material continuously back into the process without it being handled.
- The quality of the shredder or granulator size-reduction equipment affects the quality of the regrind, particularly the design and sharpness of the machine's knives, which should always be in good condition to cut the bulk material into consistent pieces.
- The melt flow of the regrind/virgin material blend may differ from what is expected, and this and other factors affect the resulting processing behavior, part shrinkage, and scrap rates.
- Heat stabilizers and antioxidants in the regrind might have been excessively consumed during its previous processing history, so much so that additional stabilizing additive levels may be needed. Thus, some knowledge about the total "heat history" of the resin (that is, the number of times it has been melted) is helpful when making decisions about incorporating reclaimed material.
- Excessive regrind or heat-damaged regrind can shift the color or affect the surface appearance or gloss of a molded

part. The clarity of transparent parts may also be influenced by regrind, changing haze values.

- In operations that produce parts in many different colors, at least some separation of scrap parts and regrind by color is necessary to produce new parts with proper coloration. This may mean complete scrap segregation by color before size-reduction, or at least separation into "light" and "dark" color streams, which then require colorants to be added to create new color-matched parts. In automotive part molding, researchers found a middle ground: an efficient "one hue dominant stream" approach, which requires separating scrap by color hue. Regrind containing at least 70% from a hue group, plus minor pigment additions, could then be used for molding a part in the proper color [1].
- Regrind (especially material with a long heat history) may reduce the strength/stiffness/ductility properties of the part or output material. A slightly greater section or film thickness in the product may even be required to prevent deformation, tearing, or cracking from regrind use (or additives added to the regrind, to enhance its properties). Accordingly, thick, bulky applications – or applications in which the regrind can be co-extruded as an internal layer of material – are more tolerant of high percentages of regrind.

Processors also commonly buy post-industrial recycled scrap from outside their facilities. These reprocessed resins can typically be bought at competitive prices from scrap brokers. They are usually supplied in size-reduced or repelletized forms and meet verifiable property and color specifications (though they can be less consistent than virgin resins, with wider-ranging properties). Although they can be called recycled, reprocessed resins are more often purchased for their prices, which may be up to 40% lower than virgin resins of equivalent properties. Such a cost-saving practice is useful, especially when using a trustworthy broker and particularly during times of high virgin resin price volatility.

7.1.2   Recycling Post-Consumer Plastic

From a life-cycle/environmental impact point of view, the recycling of post-consumer plastic is just as attractive to do as post-industrial recycling, but from a cost perspective, it is much more difficult. Whereas post-industrial scrap plastic is mainly kept within the relatively clean confines of the plant or its warehouse, post-consumer plastics experience a product's lifetime of use, exposed to all kinds of contamination, which of course makes the part harder to recycle. Nonetheless, more companies are investing in the means to recover the wealth of used plastic flowing through the marketplace. And more technologies have become available for producing cleaner recyclate more economically. These new systems are not just being adopted by recycling specialists; modular systems allow a plastics conversion company itself to create its own post-consumer recyclate.

Some processers have established their own complete recycling operations. Over the past decade, many thermoforming and bottle making companies have invested in captive recycling operations that turns old PET (bottles, edge-trim, other scrap) into usable material for the main operation. These systems can even produce material that is acceptable for food-contact packaging, with "letter of non-objection" status from U.S. regulators. Some companies have developed internal operations for recycling post-consumer polyethylene in its various forms. Such operations help close the loop for plastics, reducing the net energy required for creating new products while creating new niches for products with recycled content.

These advantages are earned not without difficulty. The mechanical recycling of post-consumer packaging (primarily of interest here is packaging made from PET and PE) has several steps, and each step presents plenty of obstacles after the recycled containers are collected. The steps described below focus on the recycling of PET containers, the largest stream of post-consumer recycled plastics, and whose recycling is a major contributor to plastics' current sustainability profile. Mechanical recycling is the most common approach for PET, though recent years have seen a boom in the growth of chemical recycling, or advanced physical recycling, where different processes such as methanolysis or solvolysis break the polymer chains back down to their monomer components. The energy required for these processes is not fully understood (or at least, not fully published) so it remains to be seen if the mass balance favors this approach over others. What follows are the more common mechanical/melt processing steps for recovering PET; similar processing steps are used to recover other plastics, such as HDPE or PP containers. This information has been codified by the Association of Plastics Recyclers in their "APR Design® Guide for Plastics Recyclability" [2].

1. **Sorting** is required to "demingle" discarded plastic containers and product streams to produce bales of post-consumer material made from mostly a single polymer and product type. Sorting may be manual or automated. Automated

sorting systems use near-infrared spectroscopy (and less commonly, x-ray sensors) to identify and divert different polymer types. This is critical because plastics such as PVC are extremely contaminating to PET in the reclamation process – with near-zero tolerance allowed (PVC, PS, and PLA products are also often hard to distinguish from PET by the eyes of human sorters).

Complicating matters is that many forms of PET products can also be incompatible with a downstream PET recycling process because of slight chemical or rheological differences. These products typically include thermoformed clamshells or trays, multilayer PET constructions, microwavable crystalline-PET trays (cPET), and glycol-modified PET (PETG). These materials are removed by manual or optical sorting. Further optical-camera or vision-system sorting by color may also be required, especially since opaque or highly pigmented containers ruin the quality of clear plastic recycling streams.

2. **Size-reduction and washing**: The mainly-PET container stream will then be ground into "dirty flake" (or "dirty regrind") using shredders or granulators (see Figure 7.1). In preliminary washing, the PET flake commonly passes through float/sink classification using a water solution in which low-density plastics, such as PE and PP cap material, floats to the surface and is removed. Automated flake sorting systems can be used at this point to separate out other polymers if an efficient container sorting method was not previously used, though these systems can be twice as expensive as automated container sorting systems.

Detergent washing is used to remove labels, glue, debris, and residue from the flake. Thus recycling processes normally use a lot of water and produce a great deal of wastewater, which must be treated for regulatory compliance before it is discharged. Low-water mechanical cleaning methods also exist, in which induced friction between flakes and centrifugal force is used to remove 80–90% of contaminants, before any water is used. Metal detectors and other systems also are used to remove metals and contaminants from the size-reduced stream, electrostatically and magnetically, and via elutriation (air-flow classification), resulting in "clean flake".

Figure 7.1 A comparison, from left to right, of recycled PET from dirty flake to rejected material to clean used flakes (Image courtesy rPlanetEarth, Inc., Los Angeles, CA).

Ultimately, the degree of cleaning required depends on the end-product destination for the recycled material. A specification for fiber grade rPET, for example, may allow PVC and aluminum contamination at twice the levels as a spec for bottle-grade rPET.

3. **Purification and reprocessing**: Clean flake is normally repelletized using extruders with screens that filter out from the melted material any contaminating particles as small as 20 microns in size. The direct use of rPET flake in processing is also possible for some applications, especially for strapping and staple fiber (for fabrics and carpeting), and non-food packaging film and thermoforming sheet.

Some processes even allow flake to be used directly in food contact applications, where reprocessing must eliminate any contaminants that affect taste, odor, or color. Such applications at least depend on passing the melted flake through automatic screen changers that are continuously cleaned, though more intense decontaminating methods are usually needed for purifying food-grade flake. A solid-state, atmospheric-pressure process has been developed that enhances the gas-phase diffusion of volatile contaminants out of the flake, without melting the rPET or excessively altering its molecular weight. Several companies have successfully used this process to create new streams of high-quality rPET for food packaging.

Along with achieving adequate decontamination, the final processing of rPET is influenced by the degree to which the polymer has degraded hydrolytically (from moisture content) and thermally (from repeated heating/melting steps). Thus, excess moisture must be removed from rPET to preserve its molecular weight profile, which is indicated and measured by its intrinsic viscosity, or IV. Various IV values may be desired for rPET; a reduced IV is allowable for rPET staple fiber production, while a high IV is helpful for producing strapping and filament yarns. For creating beverage bottle quality rPET, vacuum drying and polymer-modifying systems can raise IV from the typical rPET flake IV of 0.7–0.8 decliters/gram to the necessary level for carbonated beverage bottle production (above 0.8 dL/g).

Methods to control rPET moisture content and IV cover a range of complexities. Some or all of these might be used in a single recycling operation: controlled air/temperature storage, drying and recrystallization at elevated temperatures, vacuum degassing, solid state polymerization (SSP), and chemical additives. SSP reactor systems create conditions that branch or extend PET polymer chains, increasing molecular weight and IV. They are mainly economical for high-volume operations, while drying and vacuum systems can be modularized to suit operations with various throughputs. Meanwhile, adding polymer chain-extending additives to the recycled polymer can be a flexible approach for increasing the molecular weight and melt strength of PET and other polyesters, as well as of other recycled condensation polymers such as PC and nylons.

7.1.2.1   The Recycled Resin Challenge

Depending on what kinds of materials they are reclaiming, recycling operations must be knowledgeable about specialized equipment, raw material variation, and "tricks of the trade" to allow them to produce usable recycled resin. They face all the challenges described in detail in the last section of Chapter 2. Given the commitment required to produce good recyclate, processors are dependent on recyclers or brokers for their material – and subject to varying levels of supply, resin quality, and ever-changing price differences between virgin and recycled resin.

For obtaining a stable supply of post-consumer resin, some processors have made themselves less dependent on complicated collection, sorting, reprocessing, and brokerage networks. These companies have established their own in-house recycling operations. Such companies have invested not just in a method of controlling their sourcing of recycled material, but also in an enhanced environmental image. With the advent of the

Circular Economy, many large well-known consumer brands have made very public commitments to increasing the use of recycled content. This "demand-side" shift has led to investments in capacity for recycled materials as well as innovations in recycling technologies [3].

7.1.3   Advanced Recycling

Over the course of the past decade – and rapidly so during the past 5 years, in fact – several high-profile developments in advanced recycling have piqued the interest of the plastics industry and the public at large. As with any new, fast-moving, and dynamic market, one can only provide a snapshot at a given point in time, synthesizing and summarizing the current literature, and offering some guarded prognostications for the future. In this section, we offer a visual summary of two categories of advanced recycling (as illustrated in Figure 7.2) along with brief summaries of some of the new technologies being explored. This is neither comprehensive nor exhaustive since events are moving rapidly and simultaneously in North America, Europe, and Asia.

7.1.3.1   Dissolution ("Advanced Physical Recycling")

The previous sections focused on mechanical recycling, a physical process whereby the composition of the polymer remains unchanged. Solvent-based purification, or dissolution, is also a physical process, though it has sometimes been categorized incorrectly as "chemical" recycling. This type of advanced recycling is currently being done at scale in Germany and has been selected by Unilever to recycle olefins from landfilled post-consumer multilayer film packaging in Indonesia [4].

Delamination of multilayer film is a low-energy mechanical recycling process in which films are first shredded, then broken down further via surfactant-based micro-emulsions. A Germany-based company, Saperatec, has been operating since 2014 and works with mixed formats such as PE/Alu, PP/Alul, or PE/PET [5].

Selective extraction processes are another class of advanced physical or mechanical recycling where chemical dissolution is applied to separate and extract polymers. This is sometimes referred to as solvolysis. Companies such as APK AG (Merseburg, Germany) and CreaCycle GmbH (Grevenbroich, Germany) have developed customized solvent formulations that selectively dissolve the desired polymers so they can be recovered and re-precipitated and/or re-pelletized. The solvents themselves are recycled and the purified

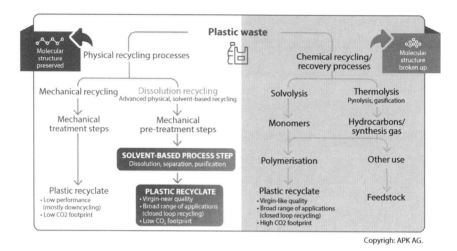

Figure 7.2 Process infographic illustrating different types of advanced plastic recycling (Source: APK AG).

polymer recyclates display properties similar to virgin polymers [5]. In North America, Procter & Gamble have invested in a high-pressure, super-critical extraction and filtration process for recycling PP. Named PureCycle Technologies, the company has developed a series of off-take agreements for "ultra pure PP" that reportedly exhibits properties identical to virgin PP. As a major user of PP in its consumer product portfolio, P&G has sent a strong signal to investors, technologists, and regulators that demand for recycled materials can be created and satisfied. It is also proof that elegant and prof-itable business models can be combined with cutting-edge polymer science.

The sub-sections below attempt to classify and differentiate chemical recovery and chemical recycling, whereby the basic polymeric structure is altered, yielding smaller molecules (monomers) from which plastics can be re-produced via polymerization [6].

7.1.3.2   Depolymerization ("Chemical or Molecular Recycling")

Breaking polymers down (or "unzipping" them) into monomers so that they can be re-polymerized is the essence of chemical recycling. Sometimes referred to as molecular recycling, and sometimes classified as a decomposition method, depolymerization of PET, for example, results in separation of its component parts monoethylene glycol (MEG) and dimethyl terephthalate (DMT). In the case of polystyrene, the depolymerization of the polymer to the styrene monomer has been shown to reduce GHG emissions by 50% when compared to using virgin, petrochemical-based styrene.

As mentioned elsewhere, several companies including Agilyx (for PS), Ioniqa, Carbios, and others have made significant steps toward commercialization. Though they differ in terms of polymers/monomers used, their business models share an important attribute: securing off-take agreements, i.e. purchasing contracts. This demand-side security ensures continued investment so that product can be manufactured and sold on a contractual basis.

7.1.3.3   Gasification/Pyrolysis ("Chemical or Feedstock Recovery")

Not to be confused with incineration, gasification or pyrolysis typically exhibit low GHG emissions, do not produce dioxins, and yield useful chemicals that can be converted back into new materials [5]. And though some people are quick to point out that these processes are not "closed loop" in the way that depolymerization is, it is important to recognize anaerobic thermal gasification or plasma pyrolysis of waste materials can displace virgin components when producing chemicals or fuels. In other words, just because petrochemical-based polymeric materials are being converted into other petrochemical-based materials, it doesn't mean that environmental benefits cannot be identified and acknowledged.

There are many pilot projects around the world where mixed plastics and waste chemicals are being converted to base fuels, gases, and waxes. Fluidized-bed gasification, for example, is where carbon-rich waste feedstocks are first converted into synthetic gas (syngas), then through further catalytic conversion and purification, refined into ultra-clean biofuels (methanol, ethanol) and chemicals. This process, mastered by Enerkem (Montreal, QC), replaces the use of fossil sources, avoids incineration-related emissions and, when compared to landfilling, re-uses carbon that would ultimately be lost or buried [7].

End-of-life options for low-value or mixed plastics are extremely limited ("bury or burn") and there is no one-size-fits-all solution. Tight regulation of landfill space in Europe, for example, leads to higher acceptance of incineration. In the United States, the opposite is true: an abundance of land, cheap fuel, and low tipping fees mean that landfilling waste materials is the path of least resistance in terms of end-of-life management.

7.2 Optimizing Plastics Processes for Sustainability

The remainder of this chapter will highlight technologies plastics processers are using to save process energy and water. These efforts optimize the basic environmental footprint or character of an operation, while creating a leaner operation that optimizes profitability. Making sustainable improvements requires first identifying the true levels of water or energy used in different plastics applications, then introducing appropriate technologies for making changes.

7.2.1 Optimizing Water Use

As shown above, plastics recycling processes use large amounts of water for cleaning reclaimed material. But in primary resin production and conversion, water usage is sometimes overlooked, or is seen as an unavoidable monetary cost and environmental issue. But water use can usually be reduced if carefully examined.

Water is required at every stage of resin production, whether making traditional synthetic resins or bioresins. At the raw material stage, water is used in oil and natural gas production, in both traditional methods and in hydraulic fracturing methods for obtaining natural gas. Water is also used in oil refining and hydrocarbon separation and monomer production. For bioresins, water is needed to grow the source crop or other bio-mass, and for the initial treatment and fermentation of the plant material. Likewise, the bacteria that produce PHAs, and obviously the algae that are being developed for polymer feedstock production, also require water. And in polymerization plants of all kinds, water is consumed.

It is hard to imagine any polymer production process in which water is not important. In resin production, water is used in closed-loop cooling systems, for the rinsing of reactors, and for stripping contaminants from residues and liquids flushed out of the process. Flocculation, separation, and clarification treatments are then required for clearing the water of wastes or reusable components. In plant-based bioresin production, wastewater containing agricultural residues such as phosphates and nitrates can disrupt the ecosystems of natural waterways if discharged.

Some polymers require more water than others in their production; given all the processing factors involved, estimates about water use range widely. In 2010, Biron [8] reported data that estimated the "direct" water consumption of polyolefins as ranging from 1.2–6.5 m^3 per ton of resin produced (a range that did not include the "indirect" water consumed by feedstock production, a consumption rate that can be even higher). A decade later, estimates reported for producing other resins range from 3–12 m^3 per ton of PVC, 1–13 m^3 per ton of unsaturated polyester, and 2–160 m^3 per ton of nylon, depending on the nylon and process type. Other studies show similar ranges, where the total water required to produce one metric ton of starch-based packaging products is estimated at 5.8 m^3. The average water use required to produce one metric ton of synthetic polymer varies

Lines, Circles, and Paradoxes

The Jevons Paradox occurs when technological progress increases the efficiency of a resource, but the rate of use of that resource also increases. It was developed in the industrial age when Britain's use of coal increased dramatically with the advent of the steam engine. In modern times, it is perhaps most commonly associated with energy efficiency: the more efficient the item (fridge, building, car), the more people use it. As applied to plastics, it can be seen in the exponential growth of resource-efficient lightweight materials. Because packages have been made lighter, more of them can be produced and transported than other, heavier materials or previous iterations of plastic items. From an LCA perspective, therefore, plastics are beneficial. What LCAs do not always consider, however, is end of life management. Thus a steep curve of growth is not matched by a flatter line of infrastructure development. Hence, the birth of the Circular Economy and the idea that industrial ecology should move from a linear to a circular system where waste is designed out and products remain in use.

Plastics are critical to automobile mileage standards, reducing vehicle weights, and therefore reducing fuel consumption. Like the fuel itself, plastics are primarily fossil-based which raises questions about overall consumption of oil- or natural-gas-based products. In countries where fuel is cheap or where other transit options are scarce, the demand for automobile-centric infrastructure remains high. This is the type of paradox that makes it clear that solutions to environmental challenges are not simple and must include a variety of stakeholders from private and public spheres of influence.

from 14.2 m³ (LLDPE) to 51.8 m³ (PS) per metric ton of plastic packaging product [9].

Of course, a key measure is the net water used for production and processing – that is, the water that cannot be recovered through treatment and closed-loop recycling. Net water use can be relatively low for some operations, but sometimes economics dictates that used water is simply treated to remove solids and chemicals as per regulated wastewater standards, and then discharged from the plant. Here, discharged water can be considered as only partially recycled by natural or man-made processes outside of a closed-loop recycling system in a plant.

In all resin production, sustainability requires using efficient methods to reduce the total and net volume of water used. Maximizing wastewater recycling must be coupled with preventing contaminated water from reaching the environment. Optimized treatment plants allow increased recycling, reducing water use. A vinyl plant in Australia, for example, reportedly reduced its total water usage per ton of product by 50% by increasing its water recovery rate to over 70%. This was after the company established a new water treatment plant incorporating strainers and a hydrocyclone and microfiltration system to remove suspended solids from wastewater, plus reverse osmosis to remove dissolved solids. The $5 million capital investment required for the plant is reportedly allowing the company to reuse water in its boiler and cooling tower and as PVC production charge water. Clearly, this is a case of a company making a relatively major investment and commitment in sustainable water use [4]. But it does indicate that practical state-of-the-art technologies are available for resin producers, and for all plastics operations seeking to optimize water use. More recently, suppliers of energy-efficient water systems have directly targeted the plastics industry as efficiency gains can attract the attention of a CFO when the return on investment offers a good use of capital. Closed-loop fluid coolers, for example, can provide savings of up to 95% of cooling water when compared to an open cooling tower [10].

7.2.2 Optimizing Energy Consumption

Energy, usually in the form of electricity, is the key consumable when converting resin into plastic products. Energy is required for drying resin and melting or softening it for molding or thermoforming, and then for cooling down the resulting hot product, typically using closed-loop chilled-water systems. There are ways of reducing the energy footprint of plastics operations – some requiring much monetary investment, some relatively little.

In raw economic terms, it is true that resin costs can be around ten times higher than the cost of the power needed to convert resin into products, with estimates of energy costs accounting for only about 3–8% of production costs. Of course, these figures will change from country to country (and even from state to state within the US where prices vary). But unlike volatile resin prices, energy consumption is in the control of plastics processors themselves. Thus, some processors are finding energy savings to be a "low hanging fruit" to increase profitability, though many processors may be ignoring these opportunities. The cost reductions of such improvements can be significant. For example, plastics industry companies have been able to save up to 10% in yearly energy costs by following recommendations from the U.S. Department of Energy sponsored Industrial Assessment Centers [11].

Of course, in sustainability terms, reduced energy consumption means reduced fossil fuel use and greenhouse gas emissions – achievements that processor companies can use to improve their reputation. With combined energy saving technologies, energy consumption per part produced in injection molding, for instance, can be cut up to 50% or more, translating into reasonable payback periods that can be used for investment into upgraded or new equipment. Various means for reducing this processing energy are overviewed below, whether they come from additional controls, the refurbishment of existing production equipment, or from the newest systems and technologies.

7.2.2.1   Refurbishing Equipment for Energy Savings

Apart from basic energy-saving ideas that apply to all kinds of manufacturing (such as high-efficiency lighting), many areas of plastics processing can be targeted for energy reduction. Anywhere heat is added to or removed from resin are possible opportunities, as are machine motions in which large masses of metal are moved continually, like in injection molding presses and screw processing. Computer process controls can be used to limit the amount of heating or cooling in a process to only the essential amount that is needed. Some concepts and examples of other technologies include:

- **"Hot feed" material handling.** In large resin production operations, raw polymer that is still hot from polymerization can be transferred to the compounding operation before it has a chance to cool down, avoiding potentially around 60°C of temperature loss from each batch [12].

- *Bulk storage pellet drying.* In this practice in conversion facilities, pellets are dried while they are in bulk storage rather than in smaller batches on the production floor, resulting in net energy savings.
- *Computer-controlled resin drying.* These systems eliminate excessive energy use through closed-loop monitoring. Resin dryers and other auxiliary equipment can be linked with primary process machines via a thermal monitoring system, which adjusts temperatures in the process to optimize production efficiency and product quality [13].
- *Process software for energy optimization.* Software that takes resin properties into account can identify optimum temperatures, mechanical energy inputs, or mold residence times for a given part and material [13]. New process machine-based software can display real-time energy use, allowing operators to modify settings for maximum efficiency, and allowing the company to monitor the performance of its entire battery of machines.
- *Optimized screw designs or reconditioned screws.* These can melt and pump molten resin more efficiently, optimizing the use of mechanical and heat energy. In a related way, *processing lubricants* can also enhance the flow of the melt, increasing through-put while decreasing screw torque, barrel temperature, and downstream cooling requirements (though the additives themselves increase material costs and can affect product properties) [12].
- *Insulation.* Covering process equipment and piping, insulation seems like an obvious energy saving approach. Insulation of the screw barrel can minimize the heating required in screw processing, reportedly resulting in a 20–40% energy savings. This translates into a return-on-investment of less than two years, according to estimates made, for example, concerning KraussMaffei's EcoPac insulation system.
- *Water recycling cooling systems.* Automatically recycling cooling water can save thousands of kilowatt hours per month in large operations, although the systems' payback periods are more easily measured in years. Chiller water temperature can also be monitored and regulated better with advanced controls, minimizing chiller loads. And water returning from the process can also be "air-blast" pre-cooled

with ambient air of low relative temperature, saving some energy.

- **Sensors.** Used for monitoring process temperatures, sensors improve product quality while helping to minimize energy use. For example, in thermoforming, infrared sensors monitoring the sheet's surface temperature help prevent overheating of the material. This helps reduce start-up times, reduces scrap, and ultimately allows the process to be set to its fastest possible speed – meaning less energy expended per unit product produced.

- **Compressed air recycling.** Compressed air use can be optimized by repairing leakage and installing better controls on smaller compressors – and by higher-tech solutions as well. For example, of importance particularly in blow molding operations, compressed-air storage systems can reduce energy by minimizing the amount of time the plant's air compressors are running. The systems allow nearly half of the air and air pressure used in blow molding essentially to be stored and recycled. Payback periods for the system are estimated at up to one year when molding half-liter bottles, and shorter periods when molding larger bottles.

Other benefits linked to certain energy-saving process improvements include improved product quality, reduced scrap and start-up times, decreased cycle times, and increased throughput. The investment costs can be high for making some of these improvements, though companies have found they can pay them off with measurable, long-term benefits. As ancillary equipment becomes more efficient, and local utility programs continue to offer incentive programs for audits and certain types of equipment and infrastructure (lights, compressors, etc.), more plastics processors are finding that these decisions are simply good business practices.

Many of the above ideas can be integrated through a basic refurbishment of a machine's components and/or systems or through structured maintenance programs. New machines themselves, however, have so many energy optimization features built-in that they are hard to ignore.

7.2.3   Choosing New Machinery for Sustainability

Primary plastics process machinery has become more energy-efficient. The trend towards green machines has spread worldwide, including China, which has targeted particularly strict standards for injection molding press efficiency. These standards and third-party verification are used for certifying when a new injection molding machine can be labeled as energy efficient, yet different materials process differently, meaning it is not as simple as generating a type of EnergyStar label for equipment.

As examples of how new primary machines are meeting these energy-efficiency goals, the newest injection molding machine technologies are summarized below.

All-electric injection molding presses have become more popular since the 1990s. All-electric machines' movements are directly driven by servo-electric motors. This is unlike with traditional hydraulic molding presses, which use stored hydraulic energy as the basis for driving machine movements. The energy use of "all-electrics" can deliver 50-70% of that required by hydraulic presses [14], and they are cleaner to operate, making them a good fit for electronics and medical molding. All-electrics' prices are coming down, but for companies molding common products, their energy savings is still often not high enough to justify the purchase of an all-electric press, a phenomenon that perhaps underscores how financial sustainability is not the same as environmental sustainability.

Hybrid or servo-driven machines can be found in several different plastic processing sectors including injection molding, extrusion, and thermoforming. Though the pace of innovation has been uneven across these processes, the movement away from pneumatics and hydraulics is unmistakable. The vast majority of machinery today runs on servo-driven platforms, primarily in chain indexing, press movements, and part removal systems. A servo drive generates energy during braking since it works like a generator. Usually, this braking energy is discharged to the surroundings as heat. Feed-back drive technology means the energy generated by the brakes flows into the intermediate circuit storage. The drive controllers are connected to this circuit, allowing the energy to be used for a different servo drive [15].

7.2.4   Sourcing Options for "Green" Energy

Some plastics companies have taken a more radical approach for controlling costs and maximizing sustainability. They produce their own process energy on-site or near the production plant. The approach echoes that in Henry Ford's largest automobile plants, which had power plants on-site that supplied all the plant's needs. But in the twenty-first century, the energy produced on-site must be green as well as affordable. Reducing their fossil fuel energy footprint using renewably produced on-site electricity allows plastics processors to claim gains in sustainability that they might have more difficulty in achieving through material or product changes.

The last decade shows that captive renewable energy projects are coming online at plastics processors willing to take on the initial investment. Direct solar panels, solar farms, and battery storage projects are becoming more popular as companies take advantage of state and federal incentives. Merrill's Packaging, a thermoforming company based in Burlingame, CA, provides neat snapshot of how industrial facilities can adapt to sustainability mandates, whether they come through regulations or customer demands. The company is audited for its water and power usage and over a four-year period, while business increased by 20%, their power use stayed the same.

Merrill's has installed a battery pack system that is charged overnight and sits in reserve. The system was supplied by STEM, an AI-powered energy intelligence firm that helps companies reduce energy consumption through smart meters and virtual energy storage systems. To reduce their impact on the community's power grid during peak use periods (typically 2 to 6 pm), the facility pulls power from the battery pack system, which reduces the amount of energy used from the grid. The system is then "refilled in the middle of the night," when power use in the community is lower [16].

Achievements like these will become more common especially when they are financially sustainable. Thus a continuing theme in this chapter was *costs* – monetary savings – required to justify the use of the greenest technology. However, the appeal of sustainable processing is also seductive and can be marketed to improve the environmental profile of a company, bringing it respect and allowing it to improve its market share, while acting as a counterweight to criticisms about uses of plastics that are not viewed as sustainable.

References

1. Garrett, Jr., D.L., Bai, H., Gu, J., Gupta, U., Kresta, J.E., Sendijarevic, V., Klempner, D., Recycling of mixed color automotive thermoplastics (SAE Technical Paper 981155). *Soc. Automot. Eng.*, 1998.
2. https://plasticsrecycling.org/apr-design-guide/apr-design-guide-home
3. https://plasticsrecycling.org/recycling-demand-champions
4. https://asia.nikkei.com/Business/Unilever-to-test-new-packaging-recycling-tech-in-Indonesia

5. Cooper, T., Overview of Developments and Innovations in End-of-Life Technologies for Flexible Packaging. *Society of Plastics Engineers Recycling Division Virtual Conference*, 2020, September.

6. *The Role of Chemistry in Plastics Recycling*, pp. 34–37, Carl Hanser Publications, Kunststoffe International, May, 2020.

7. https://enerkem.com/process-technology/technology-comparison/

8. Biron, M., Water footprint: The next challenge for the plastics industry. *SpecialChem*, http://www.specialchem4polymers. com/resources/articles/article. aspx?id=4671, 2010, July 5.

9. Rudnik, E., Environmental Impacts of Compostable Polymer Materials, in: *Compostable Polymer Materials*, 2nd Edition, Elsevier, 2019.

10. Fosco, Amsterdam, A., Close the Loop to Save Water & Energy. *Plast. Eng.*, 72, 3, 40–43, 2016, March.

11. https://www.energy.gov/eere/amo/advanced-manufacturing-office

12. Tolinski, M., Practical energy savings. *Plast. Eng.*, 63, 12, 6–8, 2007, December.

13. Toensmeier, P.A., NPE exhibitors tout sustainability. *Plast. Eng.*, 65, 9, 6–9, 2009, October.

14. https://www.plasticsnews.com/injection-molding/electric-injection-molding-presses-efficiency-key

15. Carlin, C., Five Big Advances to Track in Thin-Gauge Thermoforming. *Plast. Technol.*, 42–45, 2020, June.

16. Thermoforming Company Embodies Triple Bottom Line with Environmental Certifications, in: *SPE Thermoforming Quarterly*, vol. 37, no. 4, 2018.

8

Conclusion: Grey is the New Green

Since the first edition of this book, concerns about global warming and fossil fuel consumption have continued, with more specific concerns about the impact of plastics on the environment. As event managers can attest, an interested party could attend a sustainability-themed conference or webinar every week of the year. There has also been tremendous growth in sustainable brands, eco-conscious cleaning products, plant-based packaging – not to mention a relatively new movement toward plant-based diets. Governmental policies are being implemented to discourage use of certain materials and to encourage the use of others. Recycling is under attack in some quarters for not being efficient enough, yet market forces are stacked against some types of closed loop business models. At the time of writing, oil prices have risen to almost $40/bbl when in April of 2020 they actually traded at negative values. A decade of investment in chemical refining capacity has kept virgin plastics at historic lows, creating a sustained trade surplus for the US. New and better green technologies are

Michael Tolinski and Conor P. Carlin. Plastics and Sustainability 2nd Edition: Grey is the New Green: Exploring the Nuances and Complexities of Modern Plastics, (225–254) © 2021 Scrivener Publishing LLC

being commercialized, and more are seeing increased levels of investment. Where is the path towards sustainable plastics likely to lead? We already know, and we are living in the midst of it.

This chapter will not attempt to predict the future, but it will discuss what recent trends seem to be signaling the next few turns in the road related to plastics production, use, and sustainability. Investigated topics include:

- Trends affecting future global plastics use (8.1):
  - Consumer needs and market growth
  - Fossil fuel availability and prices
  - Alternative feedstock trends
  - Industry's priorities in responding to calls for sustainability
  - Plastic bans and controversies.
- Future progress in promoting plastics sustainability (8.2):
  - Improved partnerships, standards, industry practices, and public education
  - New sustainability-enhancing uses of both fossil- and bio-based plastics
  - From R&D to real world: Newer, more renewably-based polymeric materials.

The economic and political developments influencing sustainability discussions were still rapidly shifting as of mid-2020, making the target of "the future" a difficult one to aim at, and impossible to hit. Still, certain conclusions and core assumptions about plastics use in the future can be made, based on recent reports, data, and demographic trends – both global trends and trends related to plastics use.

8.1   Trends Affecting Future Global Plastics Use

In the production and consumption patterns of plastics, as in all modern industrial economics, fundamental issues of supply and demand apply – but what determines demand, and what affects supply? The answers to these questions become more interesting when life-cycle and environmental sustainability issues are thrown into the equation, not to mention the dimly understood notion of cost externalities [see sidebar on pp. 231]. The demand for plastic products will not only be affected by growing populations in developing countries (which are slowly becoming richer), but also by consumer attitudes and values, and by changes in buying habits, laws, and regulations. And the supply of inexpensive plastics will of course be limited by the price and availability of fossil fuels for the time being, and eventually by the price and availability of raw plastic feedstocks produced by agricultural or biological means.

The ways in which various factors underlying supply and demand interact will define the next few decades as being a truly post-industrial age for plastics. If environmental concerns become factored into more buying and manufacturing decisions made by a new, younger generation of consumers/producers, traditional ways of tracking trends in plastics demand may no longer be adequate. With these people, an idea that could be basically described as "We must try new ways of looking at things and doing things" is becoming a more common refrain, as society learns more about the environmental impacts of technology-driven lifestyles. As a result, the true costs of using often ignored plastic materials will be further exposed and debated.

8.1.1   Consumer Needs and Market Growth

The issue of consumer needs has always been a controversial one for plastics, since one person's "need" is not another's. More and more societies, though, have now become dependent on the many uses of plastics. Polymer-based products have become critical for large-scale infrastructure uses mostly outside of the consumer marketplace (for example, as landfill liners, agricultural films, piping, and flame-resistant wire and cable coverings, to name just a few). Some plastic products have become necessities even in very poor communities (consider the humble plastic pail for hauling water), and plastics have become essentially the only material option used for some sterile medical applications that save lives.

However, *consumer* needs for plastic products have simply been defined by consumers' buying choices when they recognize the value certain plastic products bring to their lives. Many people would argue that these are not needs at all, but this remains beside the point; overall, forces driving the global market are creating an increasing demand for plastic products by an increasing number of people who can afford them.

But the next phase of plastic demand growth will be different than the one that occurred in western countries after World War II. During most of that period, plastics were seen as cheap, throw-away commodities, not differentiated much by material type or environmental footprint. Now, developing countries such as China and India (with a combined population approaching 3 billion) are already adopting similar plastics-heavy consumption patterns. But greater knowledge about plastics sustainability issues has already shifted their patterns. Research about the effects of plastics' litter on marine life, for example, has influenced what kinds of plastics products are being allowed in coastal cities around the world. Bio-based plastic development will also allow these traditionally agriculture-based societies to develop their own sources of renewable feedstocks.

Of most interest when discussing sustainability is the most visible consumer-driven use of plastics: single-use packaging. Along with greater efforts in recycling, bio-based plastic packaging will become a more available option in more markets, though infrastructure requirements to manage end-of-life issues will have to be developed. Major brands have made very public commitments to reducing their use of virgin plastics. PET bottles, in particular, have evolved to the point where they can be produced via molecularly recycled materials. Danone's roadmap for Evian water, for example, was expected to lean heavily on a new partnership with Loop Industries, a Canada-based start-up that uses technology based on hydrolysis by solvents to depolymerize PET into dimethyl terephthalate (DMT) and monoethylene glycol (MEG). Loop had also recently signed agreements with both Coca-Cola and Pepsico.* Such products are visible to billions of people daily and will likely influence the actions of other packaged product makers.

Successful companies do not invest in such technologies without the potential of reasonable returns and marketability, though initial improvements in the environmental footprint of packaging from these developments may be small. However, demand for green packaging continues to grow alongside overall plastics packaging. Bottled water, as just one segment, accounted for the single largest volume share of global packaging in 2018. At 5-6% compound annual growth (CAGR) through 2023, the ubiquitous PET water bottle represents 83 billion units of absolute volume growth from a base of 260 billion in 2018. This is a staggering number, especially when you consider that the next closest category of PET bottles is carbonated drinks, with less than half the total units [2].

Global growth of bioplastics will remain in the double digits through the next decade with multiple research reports pointing to 18-29% CAGR [3].

---

* Loop Industries (NASDAQ: LOOP) came under close scrutiny following a report by Hindenburg Research, a forensic financial research firm. The publicly-traded firm was accused of defrauding investors, partners, and the public by manipulating laboratory test results to suggest that its process was closer to commercialization than it really was. Several of Loop's partners have dissolved agreements and the company's stock price has declined significantly. Investigations continue.

Even with greater use of biodegradable bioplastics, much still will need to be done to create systems for separating materials like PLA from the waste stream and sending them to industrial composting or recycling facilities that can process them. Such expectations for growth will also require overcoming the property limitations of PLA and other biologically synthesized polymers. It is not certain that consumers would otherwise be willing to make compromises and accept lower properties in these bio-plastics or other plastics containing renewable content, at higher prices as well.

Given these issues and other arguments covered in this book, certain key conclusions can be proposed about general global trends:

1. As new generations of consumers become more aware and concerned about sustainability issues, consumer markets for bio-based plastic products can be created and expanded. (And some consumers are already willing to pay more for products advertised as sustainable.)

2. The market value of biodegradable, biologically synthesized plastics such as PLA can, for now, only be supported by the renewable origins of their raw materials. Large-scale systems for the collection, industrial composting, and recycling of bioplastics are still largely undeveloped, thus most PLA is destined for landfills or incinerators as with most other plastics. Increased regionalization here means that different solutions are progressing at different rates around the world.

3. Traditional polymers made from bio-based chemical components have great potential, since the resulting materials are chemically the same as current fossil fuel grades, and their performance is likewise identical. Thus, consumer expectations can remain the same for products based on these plastics, and the material can be recycled using current recycling practices. (Such materials, however, can and should of course be advertised as bio-based, since otherwise consumers would be unaware of their origins.)

8.1.2 Fossil Fuel Availability and Price

Predictions about "peak oil" and the limits of oil production have kept oil prices and oil-consuming markets unstable since the 1970s. During this time, oil-dependent industries, such as the chemical and plastics industries, have suffered from occasional cost volatility. As noted at the beginning of the chapter, this volatility continues. Meanwhile, the risk of environmental damage when drilling for oil or gas will become greater as these resources become less accessible. This will result in more events such as the 2010 BP Gulf of Mexico Deepwater Horizon drilling disaster. Meanwhile, controversies have arisen about the potential environmental damage of using hydraulic fracturing (fracking) for extracting natural gas in the United States, for example. Fracking has been banned in some US states and European countries, yet companies like Shell have made massive investments in other states, like the Pennsylvania Petrochemicals Complex, 30 miles northwest of Pittsburgh. Even with a "clean" fossil fuel like gas, questions remain about whether a method like fracking contaminates water supplies or releases excessive methane, a potent greenhouse gas, into the atmosphere [4].

For better or for worse, there is a great deal of natural gas in the United States and Middle East that can be extracted, keeping many traditional commodity resins' prices relatively low. The flow of oil produced around the world will not likely be reduced much as new sources are brought online. The percentages of oil used for heating, transportation, and polymer production are shifting, however, so deeper industry dynamics may yet shift supply and demand further. Divestiture movements have achieved some modicum of success in the past decade [5] and the fear of owning 'stranded assets' appears to be taken seriously by some CFOs [6]. Assuming petroleum prices increase, they may only do so slowly over time. This means that bioplastics may remain at a cost disadvantage for many years, compared with traditional resins. Some other, synthesized conclusions thus follow:

1. Oil and natural gas will slowly become harder to extract, and this, along with other global/political events and trends, means that extreme periodic price swings and gradual price increases should be expected for as long as we use fossil fuels. With the costs of making most plastics being heavily dependent on fossil fuels, resin price swings will continue.
2. When certain price points for oil or gas are reached, certain bio-based plastics will start to become competitive based on cost alone, though estimates will vary on what these price points are.
3. The new methods and locations required for obtaining fossil fuels risk ever-greater environmental damage and accidents. These will result not only in natural and social costs, but also in a more damaged reputation for fossil fuel-based products such as traditional plastics. This damaged reputation may

Cost Externalities: Who Pays?

Humanity benefits from nature's bounty – clean air, water, fish stocks, fertile soil, etc. – but it is not clear if humanity adequately or accurately values these public goods, this natural capital. When these goods are damaged through private activities and pollution or overfishing is the result, who is responsible?

Pollution is usually the first example considered when explaining externalities. It is detrimental to society as a whole, but it is not taxed or "priced in" to any single product or service. Recent regulations for environmental disposal fees or increased prices for certain types of low-sulfur fuel are perhaps addressing this. Market forces, however, as explained by classical economics do not account for the asymmetric assignment of benefits and costs where pollution is concerned. Nobel Prize winner, Ronald Coase, addressed this in great detail in his seminal work, "The Problem of Social Cost" (1961): "For bargaining solutions to be feasible, property rights must be well defined, bargaining transaction costs must be low, and there must be no uncertainty or asymmetric information, when one side knows more than the other about the transaction."

Theoretical and legalistic, perhaps, but it goes to the heart of today's discussion about plastics litter and other environmental disasters ranging from Deepwater Horizon to mining spills in Russia: we cannot continue to privatize profits and socialize losses, otherwise we run the risk of creating a tragedy of the commons. If no one takes responsibility, everyone will ultimately pay the price in the form of loss of critical biodiversity than cannot be replenished.

create new market opportunities for producers of bio-based plastics.

4. Increased signs of potential damage from global warming may reduce fossil fuel use in all applications, increase the emphasis on plastics recycling, and increase the use of "carbon-neutral" bioplastics (that is, bioresins whose production processes can be shown to remove at least as much CO_2 from the atmosphere via raw materials growth as they create during resin production).

8.1.3   Alternative Feedstock Trends

The extent to which biological materials can be grown and processed into polymers efficiently will dictate the scale at which they will be used as a significant proportion of all plastics produced/used. Current bioplastics on the market are largely based on food crops such as corn and sugarcane. Given the limits on available land for growing cultivated plants, future bioplastics will require more non-crop plant sources to be used, including algae and waste biomass from plant and animal processing.

Nonetheless, most current bioresins are based on food crops, and bioresin production has not (yet) interfered with food production. Bioresins now consume only a small portion of industrial uses of crop-based starches and sugars, relative to the amount consumed by biofuel production, for example. This "food vs. fuel" debate (where "fuel" functions as a proxy for bio-based plastics because bio*fuels* such as ethanol derived from agricultural products reduce the food supply) has been hotly debated for years. In the US, renewable fuel standards (RFS) led to large investments in 1st and 2nd generation biofuels production, many of which ultimately failed when RFS subsidies were withdrawn [7].

Faced with these shifting economic and technological factors, some overarching conclusions and recommendations are proposed:

1. To meet optimal sustainability expectations, bio-based plastics should as much as possible be based on biological waste products or efficiently grown biomass rather than directly on fertilized food crops grown on cultivated land. The agricultural industry will be a key partner in increasing volumes of the desired feedstock sources.

2. When food crops such as corn and sugarcane are used as raw materials for polymer feedstock, their price volatility, depending on the growing season, can be as great as (or greater than) the price volatility of petroleum. Moreover, moral implications exist when these crops are needed to help keep food supplies at affordable price levels to feed human populations of limited means.

3. The research and development of new biomass sources for plastics (algae, plants fibers, etc.) should be encouraged, within reason. Future sources for bioplastics should come from a variety of materials, to avoid overdependence on an inflexible bio-feedstock infrastructure.

8.1.4   Industry Priorities for Sustainability

Though plastic materials are highly engineered, much of the plastics industry has a commodity focus: thus cost is king. The industry will respond to various trends over the long term, but in the short term, thin margins at resin makers and converters demand caution. This cost-caution has sometimes worked against them making changes related to sustainability. For example, apart from recent controversies about bisphenol-A and PVC, plastic resins and formulations have become safer over time, largely eliminating the use of toxic materials in additives and colorants. However, the low cost and high effectiveness of many of these materials made them so economical that they have been difficult to eliminate.

Bio-based resins and recycling developers have likewise faced special challenges that have slowed their development, sometimes nearly stopping it for decades until market forces and investors gathered to spur it. Industry news media tend to emphasize every development made related to bioplastics and recycling, thereby implying steady progress is being made. But the market penetration of these materials has been impeded by long periods of evaluation, trial runs, delayed projects, and additional research until problems can be solved and the materials are fully understood.

Bioresin developers in particular have sometimes seemed to display a "two steps forward, one step back" pattern toward commercialization. Years of development have been required to produce usable grades of bioresins at commercial quantities and low enough prices. PLA has completed most of its journey toward commercialization, though cost and supply issues remain. Bacterially produced PHAs remain in the middle phases of commercialization and still face practical hurdles related to processing temperatures, decomposition issues, and overall performance [8]. Still, there have been industrial-scale successes for companies such as Danimer Scientific (Bainbridge, GA) and Bio-on SpA (Bologna, Italy).

By comparison, the development of polyethylene from Brazilian sugarcane ethanol has seemed relatively quick. Meanwhile, one hopeful sign is that large resin makers are gradually increasing their investments in bioresin production to create larger volumes.

Large companies that use a lot of packaging plastics, like Coca-Cola and PepsiCo, want to use more bio-based and recycled plastics. But again, the correct path towards sustainability with plastics is not always clear at first. Users and producers of plastics packaging find themselves caught between meeting environmental goals and providing improved performance. Such a conflict might lead a company away from using a bio-based material, toward an improved design made from a traditional plastic. UK based supermarket chain, Tesco, has decided that PLA will not be used in any of its stores. This decision was made as part of their "traffic light" framework approach that categorizes materials into green, amber, and red buckets. PLA, along with PVC and PS, is in the red group because it cannot be adequately sorted and recycled in the current end-of-life management infrastructure [1].

These complexities and contradictions do not convert easily into conclusions; below are a few propositions based on the above section and sections from earlier on in this book:

1. Large traditional resin makers will take on greater roles in commercializing bio-based plastics, whether they are biopolymers that are biologically synthesized from renewable materials; traditional plastics synthesized from bio-based monomers; bio-based fillers and reinforcements; or blends of bio- and non-bio-based polymers.

2. Biologically synthesized resins such as PLA and PHAs face significant obstacles to widespread commercialization that are probably not emphasized enough in industry press reporting and literature. Yet their promise continues to attract investment and scientific talent.

3. Except for their lack of biodegradability, traditional resins made from bio-derived feedstocks are straight-forward candidates for helping fulfill the plastics industry's goals in environmental sustainability.

4. Recycling technologies now can produce clean, food-grade resins from discarded commodity plastics, particularly PET and PE. Technologies and standard practices are limited for recycling other kinds of polymers, though specialized recyclers are enhancing their capacities for recycling discarded polymers of all kinds.

5. Advanced recycling, in particular, offers new pathways for plastics re-conversion. Though sometimes categorized as "chemical recycling", the non-traditional, non-mechanical methods used can be both chemical and physical processes. In 2019, Closed Loop Partners classified these advanced technologies into 3 groups: purification, decomposition, and conversion [9]. Some of these will reach commercial scale, others will not.

6. Plastics converters require defined environmental goals and standards for making sustainable choices with plastics.

The best green choices are often not clear cut and are often limited by the *economic* sustainability goals of the company.

8.1.5   Plastic Bans and Controversies

As advances in bio-based plastics and recycling have been made, there seems to be increasing consumer skepticism about the usefulness, safety, and appropriateness of plastic products. In terms of toxicity, all industrial chemicals that are used in multiple products are subject to the public's scrutiny (or at its worst, fear bordering on paranoia). These concerns have their place, and have rid modern lifestyles of obvious pollutants such as lead paint and most ozone-depleting chemicals. But the effects of most chemicals are unknown at the concentrations in which the public is exposed to them. And the effects are particularly hard to determine for plastic materials, which are mainly solid materials emitting only very small amounts of mobile molecules, at most.

However, unknowns can sometimes create more fear than obvious, known threats. This results in mainstream media reports that overwhelm readers with selected information about potential chemical threats. Mainstream journal articles, social media postings, and books are often written by writers with little background in chemistry, and these media reports are often oversimplified, overgeneralized, or down-right inaccurate – but they will continue to be written.

But plastic products do have real and visible effects on the natural environment. People do litter, and plastic litter ends up in waterways and in the bodies of marine life. Certain plastics do produce toxic and/or persistent chemical pollutants when incinerated, though most remain inert, even when heated. It is true that disposing a plastic product into a landfill after a brief use-life is wasteful, and many people simply have negative feelings about the persistence of plastics in landfills and the environment. (Many items persist in landfills where degradation does not occur in the absence of oxygen.) But ignorance, hatred, and fear about a group of materials so relied on in modern society will not cure these ills as much as informed discourse will.

8.1.5.1   *Bag Bans*

Related to the marine litter concern, plastic carryout shopping bags and plastic straws are seen as the main culprit requiring regulation. In response, local governments in many regions, not just coastal ones, are banning the use of these bags. Bag bans or use-taxes are increasing around the world, COVID-19 disruptions notwithstanding. The entire nation of Italy banned single-use PE carrier bags at the start of 2011, and Ireland has been taxing bag use since 2002 [10]. Few in the plastics industry see this as the correct response for overall environmental sustainability, because paper bags used as replacements have a larger environmental footprint in their production (and obviously because the business of companies that produce plastic bags will be impacted by the spread of bag bans). There is also the sense that bag banning is just a way to scapegoat certain plastics, partially appeasing some of the public's negative feelings about plastics in general. However, bans are a reasonable response for a society that sees litter problems as otherwise uncontrollable.

Increased bag recycling and the use of biodegradable plastics are two technical responses to what has become a politicized issue. The post-consumer plastic film and bag collection rate is slowly increasing in the United States, but most efforts are voluntary. Even the most comprehensive reports acknowledge large data gaps when trying to account for post-consumer recycling efforts. More Recycling, for example, reported 2017 figures of approximately 1 billion pounds of plastic film was recovered, a third of which was exported [11] (this was prior to China's National Sword Policy). Difficulties in recycling collected bag film include contamination, which lowers its value and discourages its recovery. But questions remain whether biodegradable (or compostable) bags actually consume more material and thus more energy than the PE bags they replace and about whether these bags biodegrade adequately or still create problems when they end up in the natural environment.

Ultimately, a practical "technological fix" for this issue that solves all problems remains elusive. The issue will likely remain political and value-driven, with concerns about the harm of plastic litter to life in the natural environment usually winning out over complex arguments about life-cycle energy costs to produce various bag materials.

8.1.5.2 Post-Consumer Plastic Recycling

Despite consumers' dislike of plastic waste, recycling rates in many parts of the world, especially in the United States, remain low and are only slowly increasing. Apart from the technological challenges of sorting through plastic products in recycling bins and then cleaning and reprocessing them, basic container collection rates themselves are low. Overall, consumers are simply not trying very hard to recycle even the most common plastic packages. Bottle deposit laws do motivate the recycling of bottles, and more people in the plastics industry have started to change their minds and favor deposit laws as a way of increasing valuable PET recycling streams. Curbside collection programs in developed countries are common, though the quality and contamination of the plastic collected keeps the potential recycling yield of this material well below 100%.

In the United States at least, collection and recycling rates could be higher, to say the least. Beverage bottle collection rates are far below those of its neighbor Canada, the EU, and many other countries. This probably indicates differences in public ethics about recycling, something which can be addressed, but requires systems thinking with a touch of behavioral science. When comparing plastics to other materials, data show the United States' total recycling rate of plastics in the municipal waste stream (8%) falls short of the rates for paper and paperboard (66%), and glass (27%) [12]. Consumer confidence in plastics' sustainability is neither encouraged nor reflected by these figures. But the figures do help drive a perception that the U.S. plastics industry is being cynical in its push for recycling, since common industry statistics show that industry practices and economics seem to favor the use of virgin plastics.

Yet this does a disservice to the scientists and technicians who are working on both traditional mechanical and advanced recycling technologies. Since the first issue of this book, an entire industry has emerged that now sustains global conferences, books, and peer-reviewed journals. In addition to scientific interest, major brands are partnering with chemicals majors to invest billions of dollars and euros in new technologies. Advanced recycling offers intriguing potential to help deal with materials that are notoriously difficult to manage in current mechanical recycling systems. Flexible packaging, efficient as it is, has low bulk density which drives down economic incentives for the waste management industry. Because flexible packaging is often structured as multi-layer with different materials performing specific functions, it requires novel recycling methods or a complete redesign, or both. Though Dow Chemical and a few others have developed monolayer PE structures (HDPE + LLDPE with low levels of compatibilizers and barriers) that can be recycled in certain areas, it takes time and investment for an entire supply chain to switch materials.

Other related issues were covered more in depth earlier in this book, leading to some basic conclusions:

1. Public concerns about plastics' litter and the effects of industrial chemicals, including plastics, will continue. This will result in continuing calls for banning various kinds of plastic products over time. New candidates for banning will always be available in a society where such manufactured items are so prevalent and widespread (not to mention that the branding of certain material objects as "evil" or "unclean" is a theme that runs through the history of human civilization).

2. Continuing consumer concerns about the possible effects of synthetic chemicals used in plastics will shift attention toward alternative plastics or natural bio-based plastics not made with these chemicals.

3. Some plastic resin makers and converters can benefit from regulations or restrictions on certain plastics. (For example, recyclers can benefit from regulations expanding recycling, producers of non-BPA-containing resins can offer their materials as replacements for polycarbonate in the housewares market, and of course, biodegradable resin makers can offer their more natural materials as alternatives for those concerned about plastic waste.)

4. Plastics recycling and the use of bio-based/biodegradable resins probably cannot be increased to the point where all the public's concerns about plastics are satisfied, even with new technologies, extreme government mandates, and bans. Plastics in some form will always be used for as long as modern civilization continues, and plastics will probably always remain suspiciously mysterious and potentially harmful to some concerned consumers.

8.2   Future Progress in Promoting Plastics Sustainability

To overcome both the technological and social obstacles on the path toward sustainability, developers and manufacturers of polymeric materials will continue to need newer methods, materials, and perhaps even regulations. The industry has passed the point in history when a company simply adapting to basic conditions of supply and demand was once considered adequate for survival. Now, energy prices, supply pressures, and emissions, waste, and toxicity monitoring require a plastics industry that is making steady progress toward being leaner and greener, and more innovative in the way it responds to trends.

8.2.1   Improved Partnerships

Progress toward sustainability will require better collaboration between units within the plastics and chemicals industries and better communication with regulators and the public. Many important collaborations have already begun that are helping improve relevant industry standards and practices, while some groups are reaching out to a skeptical public to explain the technical side of plastics.

Special interest sectors within the industry have always united via trade organizations to protect their interests; it should be no different when responding to pressures for environmental sustainability. To a great extent, trade groups have as much to gain from enhancing sustainable practices within the industry as they do by fighting government regulations that would impact profitability. For a group of single-use plastic packaging makers, for example, creating standards that expand recycling (and lower the costs of recycled materials) would most certainly be more important than fighting a regulation that would increase the price of natural gas, a key polymer feedstock source.

There are several sustainability-related problems that better collaboration and communication are addressing, though new, unexpected obstacles may arise down the road. The solving of current problems will be critical if plastics are ever to be shown to be environmentally friendly for a critical mass of stakeholders. These key stakeholders must include both the "accelerator pedal" of the public (consumers that demand market applications), and the "brake pedal" of government (regulators that monitor or constrain what products are in the market).

8.2.1.1   Increasing Recycling Rates

Plastics are judged negatively by their low total recycling rate, which is below those of other common materials. And this is a shame, because at least in terms of energy use, it makes more sense to use recycled commodity plastics than practically every other kind of commodity material (glass, paper fiber, and metal, except for aluminum) [13]. So the plastics industry needs better ways of bringing more, cleaner, and less expensive post-consumer plastic material back into plastic product cycles. Industry trade groups and even individual packaging-using corporations can set ambitious recycling goals, but those goals cannot be met without interorganizational teamwork that addresses the political and technological roadblocks to achieving higher recycling rates.

One effort the industry has been pursuing for years in the United States is a revision of the resin number identification code standard that helps consumers and recyclers sort plastic articles for recycling. This code system, introduced in 1988 and now managed by ASTM, restricts the potential efficiency of recycling because it fails to adequately identify distinctions in material composition and manufacturing method, while falsely implying that every piece of plastic displaying a code number is being or can be recycled. Moreover, the number "7" code appears on all sorts of plastics not fitting the other six code numbers, including compostable bioplastics, making their collection and segregation for composting difficult, even when industrial composting is available. Clearly a new, clearer code system, or subcodes, or other kinds of messages on products will be needed if recycling rates are to increase. (See sidebar on pages 242–243, "Arrows vs. Triangles".)

Similarly, specifications must be better defined for resin composition and allowable contamination in bales of post-consumer plastic. Increased bale contamination has likely resulted from expanded curbside collection practices that accept virtually all numbered plastic products. Bales containing stray unsorted materials can increase problems in the recycling process – and costs.

Any improvement in recycled-material bale standards should pay off, because there is overall growing capacity for processing recycled plastics, especially PET. Meanwhile, shifting virgin resin prices complicate the equation – sometimes increasing, making recycled resin look more attractive – sometimes decreasing to near the price of recycled resin, making recycling less competitive.

Therefore, greater supplies of post-consumer recyclable material will be needed. This means processes must be improved to accept more resin types and grades, whether the products are thermoformed, injection molded, blow molded, or even blown or cast film. Collection methods and sorting technologies must be improved to process materials of more varied compositions and contamination levels. Producers and converters can contribute to infrastructure programs and pilot projects that enable more scalable technological solutions. Though extended producer responsibility (EPR) remains contentious in the US, it is being implemented in other countries around the world without bankrupting companies. And, perhaps most importantly, consumers must be encouraged to recycle more commonly recycled products, such as PET bottles, through expanded bottle deposit laws and anti-littering campaigns.

8.2.1.2   Plastic Litter: Minimizing the Damage

Littering may be impossible to eliminate in human societies that have access to an abundance of materials and goods. In the 1970s in the United States, public education campaigns targeted reducing litter of all kinds. Now, the response seems more focused on banning the use of certain products, such as plastic bags that often become litter. Current attention is focused more on the effects of visible marine litter (mainly plastic, mainly from discarded fishing nets) and its damage to sea life. Because human behavior and the recycling of these plastics are thus far ineffective at preventing this litter, the source of the litter – the product itself – is being eliminated.

Industry trade groups are uniting to play their role in reducing this litter in ways other than product bans. These groups are calling for more collection points for recycling plastic, stiffer littering law enforcement, and other stewardship and public education programs. These aims sound helpful, though they still depend on human behavior (and politics) for implementation.

Another approach would be to make often-littered products innocuous when littered. Quickly and thoroughly degrading polymers – which biodegrade in multiple environments completely into CO_2 and water – are the ideal solution; anything with less-than-total biodegradability will inevitably create concerns about marine litter similar to what we have now.

Otherwise, the industry should expect more bag and straw ban efforts, and attempt to capitalize on the use of reusable plastic products, such as fiber carrier bags, which themselves are often made from polymer fiber. Just as when faced with the shift from photography based on physical film to digital photography, the industry should prepare itself for a shift away from traditional single-use packaging towards its alternatives, COVID-19 era rules notwithstanding.

8.2.1.3 Educating the Public about Plastics and Sustainability

When a plastics-related controversy arises, often plastics industry trade groups and companies have responded to public criticism in two ways. One way is to emphasize complicated science-based arguments that show the "real" harmless effects or favorable value of the plastic in question. The other way, becoming more common, is to redirect the argument toward efforts the industry or company has made toward sustainability. (Sometimes these efforts are major developments, though often they are only minor efforts in reducing environmental impacts, making this strategy a kind of green-washing.) But these industry responses are more akin to a legal defense strategy than a way of educating the public. And being on the defense displays a kind of vulnerability, a weakness; being in the role of an educator is a position of strength.

For example, the public should be better educated on how to identify which plastics can and cannot be recycled. A revamped resin identification code system in the United States (see above) would aid in this education by

Arrows vs. Triangles: Good Intentions Gone Awry

Increased public awareness of plastics and the environment has led to a search for more information about how best to manage these materials at the end of their useful lives. Plastics, polymers, and chemistry are complex, with many branches of knowledge spreading out in different directions. Most people who are not scientists tend to interact with plastics in a utilitarian way – the materials are ubiquitous and useful. In our consumption-driven world, however, plastics can pose problems when they are no longer useful to us. This note is not meant to address waste management, sustainable materials management, or plastics recycling. The goal is to highlight a single, small element to address a larger point: the confusion created by the numbers found on most plastic packaging.

Contrary to popular (though perhaps waning) belief, these numbers are **not** recycling codes. They are *resin identification codes*. They do not imply recyclability, though many people assume that they do, causing a chain reaction of problems. In 1988, the plastics industry trade association (then known as SPI, or Society of the Plastics Industry), introduced resin identification codes (RIC) to help materials recovery facilities (MRFs) and recycling facilities sort different types of plastic resin. The "chasing arrows" logo was created with numbers to identify these resins. Prior to 1988, plastic items were not marked or stamped with any identification. In response to rising costs associated with tipping fees at landfills, the plastics industry attempted to create a system that would allow waste management groups to segregate potentially useful, valuable materials. In the US, 39 states adopted the SPI RIC system in some form and created legislation mandating the use of the codes, though slight differences existed among states, e.g. some mandated that all items over 16 ounces required coding, while others started coding at 8 ounce items.

In 2008, the American Society for Testing and Materials (ASTM) took over management of the RIC system. This group issued new guidelines in 2010, including changing the logo from chasing arrows to a solid triangle. The numbers did not change. Because the new standard acknowledges prior regulation, it makes clear that "...existing statutes or regulation will take precedence…" over the new one. In addition, the new regulation only applies to new molds or tooling, though no enforcement mechanism is evident in the text, and modification to older items is not required. This explains why we still see chasing arrow symbols today, despite the authors' attempts to decouple the RIC system from recycling messages.

Plastic material derived from non-fossil resources such as polylactide (or PLA, which was not fully commercialized in 1998) is still classified as 7 which means it gets lumped in with such diverse materials as acrylonitrile-butadiene-styrene (or ABS, used for Lego bricks) and polycarbonate (or PC, used in many optical applications), and multilayer materials. Paradoxically, the increase in lightweight, multilayer, flexible packaging reduces overall carbon accounting, but poses new and thorny issues for end-of-life management. Some industry participants point out that today's materials are not compatible with recycling and waste management infrastructure that was mostly developed in the 1990s. Declining municipal budgets and a relative dearth of private investment in the sector have stalled greater technological innovation and improvement. In short, consumption habits – and associated waste streams are – changing faster than infrastructure systems' ability to manage that waste. Can we ask if the RIC is still relevant? What would a new system look like? Is one even required given advances in high-speed, near-infrared sorting technologies and digital watermark systems?

We don't all get updates from ASTM in our email inboxes, so it requires some effort to stay current with a topic that overlaps industry, technology, and politics. Yet we must acknowledge that convenience can lead to thoughtlessness, and bureaucratic approaches to fast-moving societal issues are sub-optimal. Much changes with time; we are required to adapt to new rules, regulations, and realities. Little signs on the bottom of empty bottles might not seem worthy of our attention, but sometimes a small change can make a big difference.

teaching consumers about the diversity of plastic types and forms (perhaps even making young student recyclers more curious about studying plastic materials). Little can be gained in the long run by having consumers think that any plastic product imprinted with the "chasing arrows" or equivalent symbol can be or is being recycled.

At the same time, the plastics industry faces obstacles in consumer education, because plastics are complex materials to most consumers. There is also a contradiction here in that even the most seemingly simple plastic products are built from complex chemistry and processing that requires a chemistry or engineering degree to understand. Even explaining that "organic" chemistry is the basis for plastics is confusing, since most people associate the word organic with natural, chemical-free products. Most consumers cannot be expected to understand subtleties on why polyethylene is used for their milk jug, while polyethylene terephthalate is used for their soda bottle – or even how blow molding and thermoforming are different. The foreign sounding words may immediately discourage them from learning anything, as soon as these complex distinctions start to be explained.

So consumers should be exposed to minimal chemical terminology and clear language about their plastic products. Common polymer abbreviations are useful but not excessively technical ways of making distinctions between plastics for consumers. Various messages, such as in signs or product labels, can be used to communicate with consumers about the demonstrable green content of packaging. For instance:

- If a package has recycled content, a label with a clear statement such as "Contains 50% recycled PET from beverage bottles" enhances education and promotes recycling, while also selling the product.
- "This package contains recycled plastic from roughly one normal PET beverage bottle," visualizes the message even better for the consumer.
- An added statement such as, "Your purchase of this product supports the use of more recycled PET plastic," allows the consumer to feel that he or she is a participant in the journey to sustainability.
- And a sign such as "Only 'number 1' (PET) bottles in this receptacle, please," forces consumer engagement and raises awareness and consciousness, as well as being clear.

Given the use of the abbreviation "PET" in these instances, PET will become even more recognizable as the "brand" of plastic that is most valuable for recycling. In the same way, "PLA" can become eventually more recognized as the "corn plastic" or "plant plastic," though the full term "polylactic acid" is unnecessarily technical. However, excessively long polymer abbreviations may be ineffective (unfortunately, even HDPE and

LDPE may be too confusing of a distinction to expect many consumers to learn to make).

Through messaging, marketers can also educate consumers about what their plastic container *does not* contain – while emphasizing that not all plastics contain the same things. This addresses consumer concerns stimulated by mass media reports that are often not very educational. We see this now with many non-polycarbonate bottles, including ubiquitous PET bottles, labeled with "Does not contain BPA." A non-vinyl material in a PVC-type application might be labeled "Does not contain phthalates or vinyl (PVC)," if the target application market has become sensitive about these materials. Thus marketing and public education goals merge, and simple phrases like these are already being used more and more on various products, along with the "How2Recycle" labels.

8.2.1.4   Implementing Bio-Based Materials

Given consumer interest and their potential to reduce plastics' environmental footprint, bio-based plastics will occupy growing niches of applications. However, whether the polymer is synthesized through a biological process, made from bio-based feed-stock, or contains natural fibers, good properties and low prices will still be expected – and greater volumes will be needed to push costs down.

Moreover, biologically synthesized plastics like PLA are relatively new materials with unique properties, and must be approved for use when replacing traditional polymers. Multiple approval processes can slow their introduction, and new material standards, such as those used in the automotive industry, may have to be written for them. Thus, biofeedstock-based traditional plastics such as sugarcane-based polyethylene and polypropylene may have the advantage of being quick and easy replacements that fit better into the approval processes.

Bioplastic introduction may also be slowed by a lack of clarity about whether a material really is a "bioplastic." Here, government agencies can be useful by setting a baseline for what can and cannot be labeled as "bio-based," similar to their efforts in defining which foods can be called "organic." For instance, in 2011 the United States Department of Agriculture established the Certified Biobased Product label and in 2012 the European Union created a Bioeconomy Strategy to develop policies that integrate bioeconomy considerations (agricultural-based products) into EU-level frameworks [14]. To be certified, products are first tested as per ASTM (or DE or ISO) standards to quantify their bio-based content.

Similarly, third-party independent testing organizations, such as the Biodegradable Products Institute, are useful for helping to establish specifications and harmonize standards that verify the green qualities of products. Such organizations are also useful for creating alliances between groups with related sustainability goals. But for bio-based plastics to play a significant role in plastics applications overall, they must be proven problem solvers, not just marketing ploys. They must be unequivocally shown to have lower environmental impacts than traditional plastics (and should be recyclable to some degree), and must be produced without affecting food production costs. At that point, the already increasing capacity of bio-based plastic producers can be encouraged by market forces and additional government policies.

8.2.1.5   Improving the Life-Cycle Impact of Plastics

Both biological and traditional plastics makers and users will benefit through improved methods of determining their materials' environmental impacts. Life-cycle assessment studies sometimes provide unexpected or inconsistent results, which are determined using varying LCA models, data, and assumptions. LCA is a tool that can be improved, and greater agreement on how LCAs should be performed is needed.

A more abstract – and therefore maybe even more difficult – challenge may simply be for the industry to agree on what a good definition of the term "sustainability" should be in terms of plastics. Sustainability's most generic definition is broad: "Meeting our current needs without preventing future generations for meeting their needs." Applying this definition thoroughly to plastics would require all plastic products to be 100% recyclable, or when not that, then fully biodegradable and made from renewable resources. When both of these criteria together cannot cover a plastics application, the plastic (or any other industrial material being considered) would not be used. Such an extreme application of the sustainability principle is unlikely to happen in market-driven economies, where the solution for a given problem is always expected to be found around the next bend in the road. Rather, strategic compromises and agreed on goals for recycling and biocontent rates encourage slow, steady progress.

8.2.1.6   Sustainability in the Product Development Process

Many companies have already brought sustainability factors into their product design and material selection processes. Large corporations such as the Ford Motor Company have even integrated within their product research and development organization long-term projects for creating uses of renewable-content plastics. But it is not easy to account for all the ways sustainable choices affect key factors in product planning, such as:

- higher costs of more sustainable materials (bioplastics or recycled-content materials);
- product design modifications for creating an optimized design using less material or alternative materials;
- processing modifications required to handle new, sustainable materials (adjusting to the way PLA processes, for instance); and
- product marketing and pricing that covers any extra costs of sustainability (perhaps requiring "selling the sustainability" of the product).

Material cost in particular has been the overriding factor for most applications of commodity plastics. "A lowest cost wins" methodology in product planning will make it nearly impossible for more sustainable plastics to be used in volumes high enough for them to become more competitive. By contrast, a "total systems cost" view brings multiple factors into a sustainable business model. Such a model integrates marketing knowledge and customer interests as factors in product planning – that is, if green can be sold, then green can and should be the basis of a new product. Consumer pressure, and especially pressure from large, global retailers, are already influencing how plastic products are developed in terms of sustainability.

The more these factors are considered early on in the product design process, the better. This is because at least 70% of a product's total lifetime costs are determined during the concept/design phase of development. Once the tooling has been cut and pilot production runs are made, low-cost improvements for sustainability are difficult to make. Thus, multiple fundamental questions about the product's use and market and about its design and material alternatives should be asked at the start of product development, not later on. Some useful answers might be found in published LCA results, avoiding the necessity of a company performing its own, possibly limited, LCA.

In product development meetings, a participant might even take on the role of "the skeptical environmentalist" – asking difficult, blunt questions about the plastic product's life cycle that might otherwise be ignored. Questions should be wide-ranging, concerning not just material choices and usage, but also about green wastes, as listed by Wills [15], which can be just as important to consider. These other wastes include water use, energy use, solid waste production, transportation requirements, emissions (VOCs, toxins, and greenhouse gases), and biodiversity effects. Biodiversity factors include both the potential harm of the product to living things (such as to marine life by littered plastic debris) and the effects of overharvesting natural resources (a potential issue in the overuse of food crops to create biopolymers, for example). By taking the environmentalist's concerns into account during initial planning, soon all planning team members will end up becoming experts in sustainability as well, making future product development easier.

8.2.1.7   Effective Government Regulation

Regulations are not always harmful to profits, and do not always need to be fought. For example, proper regulation (such as bottle deposit laws) can help the industry increase the availability of recycled material without seriously hurting new product demand.

Considering that regulations usually attempt to solve major problems by enforcing strict rules on a diverse industry, regulatory activities can sometimes go wrong and can be ineffective. On the other hand, given that legislation such as Europe's REACH chemical registration program is in force, meeting their requirements at least offers a company a chance to brag about achieving socially responsible goals.

Governmental structures can be useful for settling disputes or for providing at least some baseline certified status for industry to shoot for (see 8.2.1.4 above). And settling disputes in court, such as those regarding deceptive claims about plastic biodegradability, for instance, at least helps set a precedent for the industry to consider. Thus, the effectiveness of laws and regulations should be objectively evaluated, and the potential benefits of legal structures not discounted.

In sum, to be pursued effectively, all seven of these activities may require a change in mindset. What were once considered market-killing regulations, costly standards, or environmentalist harassment might now be seen as factors to integrate with business planning. This could help move plastics into a new phase in which they are less associated with wastefulness and the negative tendencies of consumer society.

8.2.2   New Sustainability-Enhancing Approaches

Besides dealing with sustainability issues from their common uses, plastics will play key roles in future sustainable technologies for serving the world's energy, transportation, and information needs. Plastic compounds, after all, are engineered materials – many with remarkable characteristics. Two application areas in particular are emerging that will require more help from polymer-based materials to reduce the total environmental footprint of the world's energy use. The sustainable contributions from these plastics offset by far the environmental impacts from producing their own raw materials. (And when speaking about these applications in the future, it might be better for the plastics industry to emphasize the importance of plastics, rather than of polymers or composites, to emphasize the continuous range of uses plastics have – and to neutralize the dichotomy between "good, high-tech polymer composites" and "bad, wasteful plastics.")

8.2.2.1   Energy-Efficient Transportation

Particularly in automotive manufacturing, plastics are allowing cars to go farther using less gasoline (or in the case of electric vehicles, less electrical energy). With greater opportunities in the developing world for more people to own their own vehicles, fuel-saving technologies are becoming even more critical.

As automotive plastics use expands, the use of bio-based polymers and reinforcements should also increase, along with more recycled-content materials and more recycling. (And given its demand for recycled packaging plastics, one might expect the developing world to be where automotive plastics recycling really advances.) New automotive uses of plastics point to some interesting possible trends:

- Metal to plastics replacement efforts continue in applications all over the vehicle. Some of the most noteworthy are the increase in carbon fiber reinforced materials. Examples include the General Motors truck bed and Ford's carbon fiber PP A pillar bracket which replaced ASA. Both of these items won SPE Automotive Innovation Awards in 2019 and 2017, respectively.
- The European Union is developing a mandate on recycled content for vehicles (included in the European Strategy for Plastics in a Circular Economy adopted on January 2018). One of the initial goals discussed was to ensure that all exterior, black functional plastics parts to be produced from recycled feedstock.
- Commercialization of biomass reinforced compounds. Examples include Rhetech's RheVision line of bio-fiber

reinforced thermoplastic compounds for which Ford Motor released a worldwide specification for rice hull filled compounds, and Chrysler Corporation released approvals for wood-filled and rice hull-filled grades.

- Bio-based polymers like Ford's approval of soy based expandable TPU foam for seat cushions, and EMS-Grivory's supply of nylon material based from castor oil [16].

These are some of the most active and obvious efforts. There are other ongoing projects to increase the circularity of materials within the automotive industry such as the Argonne National Lab's study on Automotive Shredder Residue, and the Plastics Industry Association's collaboration with the Institute of Scrap Recycling Industry to increase plastics recovery from end-of-life vehicles.

Plastics and their composites are also being exploited more in lighter aircraft and in other forms of transportation where vehicle mass determines energy consumption. Such uses of plastics are unfortunately often hidden from view – and from discussions about plastics and sustainability.

8.2.2.2   Flexible Solar-Energy Systems

Plastics will be crucial for developing new kinds of lower-cost solar energy systems. Solar energy is one of the fastest growing of all alternative energy technologies, and various polymers – some exotic, some common – can now be used for nearly every component in a photovoltaic (PV) system – from the backing substrates to the electricity-generating cell layers.

Research has focused on developing combinations of complex polymers instead of rigid silicon for the heterojunction cell that generates PV current (e.g., [17]). Such polymer cells will eventually be supported by a backing substrate made from PET rather than glass, allowing the PV system to be thermoformed into various shapes.

Plastics are also allowing PV systems to be integrated with everyday products. Dow Chemical Company is investing in the full-scale production of injection-molded solar roofing shingles. The shingles incorporate thin-film PV cells to generate power, and can be installed by the same roofers who install conventional shingles. Meanwhile, geomembrane covers with flexible PV strips are being installed in m any landfills to capture the sunlight falling over those large surface areas [18]. These examples indicate that there are many possibilities for integrating solar energy components in plastic parts already in common use.

8.2.3   New Research & Development

Apart from plastics' roles in energy-savings technologies, future research and development may allow bioplastics to be even more sustainably sourced and produced, perhaps in ways not yet imagined. For bioplastics of current interest, such as PLA, there is no reason to believe that their production and properties cannot still be radically improved with further development, just as traditional commodity polymer processes and quality were improved over the initial decades of their use. More recent technical developments for other sustainable materials indicate the tide of research is still coming in, rather than going out, with projects cycling through research labs to start-ups to pilot projects to, eventually, commercialization:

- A bio-based alternative to traditional PET, polyethylene furanoate (PEF), is based on carbohydrate biomass but can still be produced using the existing petrochemical production infrastructure and supply chains, aiding its cost competitiveness [19]. Example company: Avantium (Netherlands).
- New methods for developing and scaling PHA production continues with companies such as Danimer Scientific, Newlight Technologies, Bio-on SpA, and Mango Materials. These companies are using biofermentation and catalyst technologies to tune PHA performance such that it can meet a wide variety of applications.
- Finding new, economical uses for previously unrecyclable materials can help reduce the need for virgin conversion. A consortium of companies in Germany recently published findings showing that mixed, colored APET flakes (those not used for bottles) can be upcycled into heat-stabilized CPET trays [20].
- And perhaps the ultimate goal for polymer production is at hand: using the greenhouse gas CO_2 directly as a feedstock rather than using plants to convert it into a feedstock source. Essentially a type of photosynthesis, using CO_2 as a feedstock has gained a lot of interest over the past decade, with continued government, university, and private funding. Catalyst technology is critical because CO_2 typically has low reactivity and requires high-energy reaction partners [21]. Companies such as Novomer (acquired by Aramco in 2016) and Newlight Technologies

are using CO_2 as the basis for a new generation of bio-based materials.

These are just a few examples of work indicating that plastics developers have already progressed far along their journey toward sustainability. They indicate that the industry has at least somewhat reverted from perceived "maturity" back into a less mature, "fresh start" phase, and that now, to meet sustainability goals, plastics will need to re-evolve.

References

1. Carlin, C., The Circular Economy for Plastics is Moving Forward in Europe, but Challenges Remain, Conference Report. *Plast. Eng.*, 75, 2, 6–10, 2019, February.
2. Euromonitor Consulting, in: *Global Packaging Landscape: Growth, Trends, & Innovations*, July 2019.
3. https://www.usda.gov/media/blog/2018/09/19/new-industrial-revolution-plastics
4. Leahy, S., Fracking Boom Tied to Methane Spike in Earth's Atmosphere, in: *National Geographic Magazine*, 2019, August 15.
5. www.en.wikipedia.org/wiki/fossil_fuel_divestment, accessed July 2020.
6. https://www.reuters.com/article/us-bp-strandedassets-analysis-idUSKBN23V1ZY
7. https://www.bloomberg.com/news/articles/2020-07-09/biofuel-revolution-was-doomed-by-policy-and-investment-failures
8. Laird, K., Bioplastics: Promising but Pricey, in: *Plastics News*, 2019, February 5.
9. Accelerating Circular Supply Chains for Plastics, in: *A Landscape of Transformational Technologies that Stop Plastic Waste, Keep Materials in Play and Grow Markets*, Closed Loop Partners, 2019.
10. Italy sets precedent in European bag bans, in: *Plastics News*, http://www.plasticsnews.com, 2011, March 18.
11. 2017 National Post-Consumer Plastic Bag & Film Recycling Report, in: *More Recycling*, Sonoma, CA, July 2019.
12. www.epa.gov/facts-and-figures-about-materials-waste-and-recycling/, 2017 data.
13. Plastics and Sustainability, in: *A Valuation of Environmental, Costs, and Opportunities for Continuous Improvement*, Trucost, 2016, July.
14. https://www.european-bioplastics.org/policy/bioeconomy/
15. Wills, B., *Green Intentions*, CRC Press, Boca Raton, 2009.
16. Details taken from interview with Chris Surbrook, in: *SPE Recycling Division Past-President*, July 2020.

17. Petzold, S., Wang, C., Khazaal, A., Osswald, T., Conjugated polymer photovoltaic solar cells: Manufacturing and increasing performance. *Plast. Eng.*, 66, 6, 26–32, 2010, June.

18. https://www.hdrinc.com/au/insights/exposed-geomembrane-cover-design-simplified-design-approach

19. https://omnexus.specialchem.com/selection-guide/polyethylene-furanoate-pef-bioplastic

20. Engelmann, S., Stoye, C., Wefelmeier, C.-J., Kossman, A., Klammt, N., Wyss, J., Ganz, D., From A to C: Developments in PET, in: *SPE Thermoforming Quarterly*, vol. 38, no. 3, Aug-Sep 2019.

21. https://setis.ec.europa.eu/publications/setis-magazine/carbon-capture-utilisation-and-storage/co2-feedstock-polymers

Index

Printed and bound by CPI Group (UK) Ltd, Croydon, CR0 4YY